U.S. Department
of Transportation

**Federal Aviation
Administration**

FAA-S-8081-6C

FLIGHT INSTRUCTOR

Practical Test Standards

for

AIRPLANE

November 2006

FLIGHT STANDARDS SERVICE
Washington, DC 20591

NOTE

Material in FAA-S-8081-6C will be effective November 1, 2006. All previous editions of the Flight Instructor—Airplane Practical Test Standards will be obsolete as of this date.

Record of Major Changes
FAA-S-8081-6C

Introduction

- The Reference section has been updated to current FAA publications
- An Abbreviation section has been added
- Special Emphasis areas updated
- Practical Test Prerequisites updated
- Added Letter of Discontinuance information
- Added Aeronautical Decision Making (ADM) and Risk Management Information
- Renewal or Reinstatement of a Flight Instructor Matrix revised. Area of Operation II, added TASK L and one other TASK. Added Area of Operation IV, one TASK.
- Reference to the metric system has been eliminated

FLIGHT INSTRUCTOR AIRPLANE-MULTIENGINE

- Area of Operation VII, E TASK, added element e, course of action if selected touchdown point is going to be missed. F TASK, added element f, importance of manufacturer's recommended procedures.
- Area of Operation XI, D TASK, Accelerated Maneuver Stalls have been added.
- Area of Operation XIII, E TASK, added element e, follows manufacture's procedures.

FOREWORD

The Flight Instructor—Airplane Practical Test Standards book has been published by the Federal Aviation Administration (FAA) to establish the standards for the flight instructor certification practical tests for the airplane category and the single-engine and multiengine classes. FAA inspectors and designated pilot examiners shall conduct practical tests in compliance with these standards. Flight instructors and applicants should find these standards helpful in practical test preparation.

/s/ 9/22/2006

____

Joseph K. Tintera, Manager
Regulatory Support Division
Flight Standards Service

CONTENTS

INTRODUCTION........1

General Information........1
Practical Test Standards Concept........2
Flight Instructor Practical Test Book Description........2
Flight Instructor Practical Test Standards Description........2
Abbreviations........5
Use of the Practical Test Standards Book........6
Special Emphasis Areas........8
Practical Test Prerequisites........9
Aircraft and Equipment Required for the Practical Test........10
Flight Instructor Responsibility........11
Examiner Responsibility........12
Satisfactory Performance........13
Unsatisfactory Performance........14
Letter of Discontinuance........15
Aeronautical Decision Making and Risk Management........15
Single-Pilot Resource Management........15
Applicant's Use of Checklists........16
Use of Distractions During Practical Tests........16
Positive Exchange of Flight Controls........16
Initial Flight Instructor Certification........17
Addition of Aircraft Category and/or Class Ratings to a Flight Instructor Certificate........17
Renewal or Reinstatement of a Flight Instructor Certificate........17

CONTENTS........2-i
APPLICANT'S PRACTICAL TEST CHECKLIST........2-v
EXAMINER'S PRACTICAL TEST CHECKLIST........2-vii
ADDITIONAL RATING TASK TABLE........2-xi
RENEWAL OR REINSTATEMENT OF A FLIGHT INSTRUCTOR TABLE........2-xii

AREAS OF OPERATION:

I. FUNDAMENTALS OF INSTRUCTING........2-1
II. TECHNICAL SUBJECT AREAS........2-4
III. PREFLIGHT PREPARATION........2-11
IV. PREFLIGHT LESSON ON A MANEUVER TO BE PERFORMED IN FLIGHT........2-13
V. PREFLIGHT PROCEDURES........2-14
VI. AIRPORT AND SEAPLANE BASE OPERATIONS........2-21
VII. TAKEOFFS, LANDINGS, AND GO-AROUNDS........2-24
VIII. FUNDAMENTALS OF FLIGHT........2-33
IX. PERFORMANCE MANEUVERS........2-37
X. GROUND REFERENCE MANEUVERS........2-38
XI. SLOW FLIGHT AND STALLS........2-41
XII. BASIC INSTRUMENT MANEUVERS........2-45
XIII. EMERGENCY OPERATIONS........2-50
XIV. MULTIENGINE OPERATIONS........2-56
XV. POSTFLIGHT PROCEDURES........2-62

INTRODUCTION

General Information

The Flight Standards Service of the Federal Aviation Administration (FAA) has developed this practical test as the standard to be used by examiners[1] when conducting flight instructor airplane practical tests. Instructors are expected to address all of the elements contained in this practical test standard (PTS) when preparing applicants for practical tests. Applicants should be familiar with this PTS and refer to these standards during their training.

The FAA gratefully acknowledges the valuable assistance provided by many individuals, companies, and organizations throughout the aviation community who have contributed their time and talent in assisting with the revision of this practical test standard.

This book may be purchased from the Superintendent of Documents, U.S. Government Printing Office (GPO), Washington, DC 20402-9325, or from GPO's web site.

http://bookstore.gpo.gov

This PTS is also available for download, in pdf format, from the Flight Standards Service web site.

www.faa.gov

This PTS is published by the U.S. Department of Transportation, Federal Aviation Administration, Airman Testing Standards Branch, AFS-630, P.O. Box 25082, Oklahoma City, OK 73125.

Comments regarding this publication should be sent, in e-mail form, to the following address.

AFS630comments@faa.gov

[1] The word "examiner" denotes either the FAA inspector, FAA designated pilot examiner, or other authorized person who conducts the practical test.

Practical Test Standards Concept

Title 14 of the Code of Federal Regulations (14 CFR) part 61 specifies the AREAS OF OPERATION in which knowledge and skill shall be demonstrated by the applicant before the issuance of a flight instructor certificate with the associated category and class ratings. The CFRs provide the flexibility that permits the FAA to publish practical test standards containing the AREAS OF OPERATION and specific TASKs in which competency must be demonstrated. The FAA shall revise this book whenever it is determined that changes are needed in the interest of safety. **Adherence to the provisions of regulations and the practical test standards is mandatory for the evaluation of flight instructor applicants.**

Flight Instructor Practical Test Book Description

This book contains the practical test standards for Flight Instructor—Airplane (multiengine). Other flight instructor practical test books include:

- FAA-S-8081-7, Flight Instructor—Rotorcraft (Helicopter and Gyroplane)
- FAA-S-8081-8, Flight Instructor—Glider
- FAA-S-8081-9, Flight Instructor—Instrument (Airplane and Helicopter)

The Flight Instructor Practical Test Standards include the AREAS OF OPERATION and TASKs for the issuance of an initial flight instructor certificate, for the addition of category and/or class ratings to that certificate, and for renewal or reinstatement of a certificate or rating by a practical test.

Flight Instructor Practical Test Standards Description

AREAS OF OPERATION are phases of the practical test. In this practical test book, the first AREA OF OPERATION is Fundamentals of Instructing; the last is Postflight Procedures. The examiner may conduct the practical test in any sequence that will result in a complete and efficient test; **however, the ground portion of the practical test shall be completed prior to the flight portion.**

TASKs are titles of knowledge areas, flight procedures, or maneuvers appropriate to an AREA OF OPERATION. The abbreviation(s) within parentheses immediately following a TASK refer to the category and/or class aircraft appropriate to that TASK. The meaning of each abbreviation is as follows:

AMEL Airplane—Multiengine Land
AMES Airplane—Multiengine Sea

NOTE: When administering a test based on sections 1 and 2 of this PTS, the TASKs appropriate to the class airplane (ASEL, ASES, AMEL, or AMES) used for the test shall be included in the plan of action. The absence of a class indicates the TASK is for all classes.

NOTE is used to emphasize special considerations required in the AREA OF OPERATION or TASK.

REFERENCE identifies the publication(s) that describe(s) the TASK. Descriptions of TASKs and maneuver tolerances are not included in these standards because this information can be found in the current issue of the listed reference. Publications other than those listed may be used for references if their content conveys substantially the same meaning as the referenced publications.

These practical test standards are based on the following references:

14 CFR part 1 Definitions and Abbreviations
14 CFR part 23 Airworthiness Standards: Normal, Utility, Acrobatic, and Commuter Category Airplanes
14 CFR part 39 Airworthiness Directives
14 CFR part 43 Maintenance, Preventative Maintenance, Rebuilding, and Alteration
14 CFR part 61 Certification: Pilots and Flight Instructors
14 CFR part 67 Medical Standards and Certification
14 CFR part 91 General Operating and Flight Rules
NTSB part 830 Notification and Reporting of Aircraft Accidents and Incidents
AC 00-6 Aviation Weather
AC 00-45 Aviation Weather Services
AC 60-22 Aeronautical Decision Making
AC 60-28 English Language Skill Standards as Required by 14 CFR parts 61, 63, and 65
AC 61-65 Certification: Pilots and Flight Instructors

AC 61-67	Stall and Spin Awareness Training
AC 61-84	Role of Preflight Preparation
AC 61-94	Pilot Transition Course for Self-Launching or Powered Sailplanes (Motorgliders)
AC 61-107	Operations of Aircraft at Altitude Above 25,000 feet MSL and/or Mach Numbers (Mmo) Greater than .75
AC 90-42	Traffic Advisory Practices at Airport Without Operating Control Towers
AC 90-48	Pilots' Role in Collision Avoidance
AC 90-66	Recommended Standard Traffic Patterns for Aeronautical Operations at Airports Without Operating Control Towers
AC 91-13	Cold Weather Operation of Aircraft
AC 91-55	Reduction of Electrical System Failures Following Aircraft Engine Starting
FAA-H-8083-1	Aircraft Weight and Balance Handbook
FAA-H-8083-3	Airplane Flying Handbook
FAA-H-8083-9	Aviation Instructor's Handbook
FAA-S-8081-4	Instrument Rating Practical Test Standards
FAA-S-8081-12	Commercial Pilot Practical Test Standards
FAA-S-8081-14	Private Pilot Practical Test Standards
FAA-H-8083-15	Instrument Flying Handbook
FAA-H-8083-23	Seaplane, Skiplane, and Float/Ski Equipped Helicopter Operations Handbook
FAA-H-8083-25	Pilot's Handbook of Aeronautical Knowledge
Order 8080.6	Conduct of Airman Knowledge Tests
AC 150/5340-1	Standards for Airport Markings
AC 150/5340-18	Standards for Airport Sign Systems
AC 150/5340-30	Design and Installation Details for Airport Visual Aids
AIM	Aeronautical Information Manual
A/FD	Airport/Facility Directory
NOTAMs	Notices to Airmen
POH/AFM	Pilot Operating Handbooks and FAA-Approved Airplane Flight Manuals
USCG	Navigation Rule: International - Inland

The Objective lists the elements that must be satisfactorily performed to demonstrate competency in a TASK. The Objective includes:

1. specifically what the applicant should be able to do;
2. conditions under which the TASK is to be performed; and
3. acceptable performance standards.

The examiner determines that the applicant meets the TASK Objective through the demonstration of competency in all elements of knowledge and/or skill unless otherwise noted. The Objectives of TASKs in certain AREAS OF OPERATION, such as Fundamentals of Instructing and Technical Subjects, include only knowledge elements. Objectives of TASKs in AREAS OF OPERATION that include elements of skill, as well as knowledge, also include common errors, which the applicant shall be able to describe, recognize, analyze, and correct.

The Objective of a TASK that involves pilot skill consists of four parts. The four parts include determination that the applicant exhibits:

1. instructional knowledge of the elements of a TASK. This is accomplished through descriptions, explanations, and simulated instruction;
2. instructional knowledge of common errors related to a TASK, including their recognition, analysis, and correction;
3. the ability to demonstrate and simultaneously explain the key elements of a TASK. The TASK demonstration must be to the COMMERCIAL PILOT skill level[2]; and
4. the ability to analyze and correct common errors related to a TASK.

Abbreviations

14 CFR	Title 14 of the Code of Federal Regulations
AC	Advisory Circular
ADM	Aeronautical Decision Making
AGL	Above Ground Level
AIRMETS	Airman's Meteorological Information
AME	Airplane Multiengine
AMEL	Airplane Multiengine Land
AMES	Airplane Multiengine Sea
ASEL	Airplane Singe-Engine Land
ASES	Airplane Single-Engine Sea
ATC	Air Traffic Control
ATIS	Automatic Terminal Information Service
CFIT	Controlled Flight into Terrain
CRM	Crew Resource Management
DME	Distance Measuring Equipment
FAA	Federal Aviation Administration
FDC	Flight Data Center
FSDO	Flight Standards District Office

[2] The teaching techniques and procedures should conform to those set forth in FAA-H-8083-25, Pilot's Handbook of Aeronautical Knowledge; FAA-H-8083-9, Aviation Instructor's Handbook; FAA-H-8083-3, Airplane Flying Handbook; and FAA-H-8083-15, Instrument Flying Handbook

G	Glider
GPO	Government Printing Office
GPS	Global Positioning System
IA	Instrument Airplane
IH	Instrument Helicopter
LAHSO	Land and Hold Short Operations
LORAN	Long Range Navigation
MEL	Minimum Equipment List
Mmo	Means Maximum Operating Limit Speed
NAVAID	Navigation Aid
NDB	Non Directional Beacon
NOTAM	Notice to Airmen
NTSB	National Transportation Board
NWS	National Weather Service
PC	Proficiency Check
PTS	Practical Test Standard
RG	Rotorcraft Gyroplane
RH	Rotorcraft Helicopter
SIGMETS	Significant Meteorological Advisory
SUA	Special Use Airspace
TFR	Temporary Flight Restriction
VFR	Visual Flight Rules
VHF	Very High Frequency
VOR	Very High Frequency Omnirange
V_{MC}	Minimum Control with the Critical Engine Inoperative
V_X	Best Angle of Climb
V_Y	Best Rate of Climb
V_{SSE}	Safe Single Engine Speed
V_{YSE}	Single-Engine Best Rate of Climb

Use of the Practical Test Standards Book

The Flight Instructor Practical Test Standards are designed to evaluate competency in both knowledge and skill.

The FAA requires that all Flight Instructor practical tests be conducted in accordance with the appropriate Flight Instructor Practical Test Standards and the policies set forth in the INTRODUCTION. The flight instructor applicant must be prepared to demonstrate the ability to instruct effectively in **ALL** TASKs included in the AREAS OF OPERATION of the appropriate practical test standards, unless otherwise noted.

All of the procedures and maneuvers in the Private Pilot and Commercial Pilot Practical Test Standards have been included in the Flight Instructor Practical Test Standards. **However, the flight instructor PTS allows the examiner to select one or more TASKs in each AREA OF OPERATION therefore allowing the practical test for initial certification to be completed within a reasonable time**

frame. In certain AREAS OF OPERATION, there are **required** TASKs, which the examiner must select. These **required** TASKs are identified by **NOTES** immediately following the AREA OF OPERATION titles.

The term "instructional knowledge" means the instructor applicant is capable of using the appropriate reference to provide the "application or correlative level of knowledge" of a subject matter topic, procedure, or maneuver. It also means that the flight instructor applicant's discussions, explanations, and descriptions should follow the recommended teaching procedures and techniques explained in FAA-H-8083-9, Aviation Instructor's Handbook.

In preparation for the practical test, the examiner shall develop a written "plan of action" for each practical test. The plan of action is a tool, for the sole use of the examiner, to be used in evaluating the applicant. The plan of action need not be grammatically correct or in any formal format. The plan of action for an initial certification test shall include one or more TASKs in each AREA OF OPERATION and shall **always** include the required TASKs. The plan of action must incorporate one or more scenarios that will be used during the practical test. The examiner should try to include as many of the TASKs into the scenario portion of the test as possible, but maintain the flexibility to change due to unexpected situations as they arise and still result in an efficient and valid test. **Any TASK selected for evaluation during a practical test must be evaluated in its entirety**. If the applicant is unable to perform a TASK listed in the "plan of action" due to circumstances beyond his/her control, the examiner may substitute another TASK from the applicable AREA OF OPERATION.

The examiner is not required to follow the precise order in which the AREAS OF OPERATION and TASKs appear in this book. The examiner may change the sequence or combine TASKs with similar objectives to have an orderly and efficient flow of the practical test.

The "plan of action" for a test administered *for the addition of an aircraft category and/or class rating* to a flight instructor certificate shall include the required AREAS OF OPERATION as indicated in the table at the beginning of each section. The required TASKs appropriate to the rating(s) sought must also be included. In some instances, notes identify additional required TASKs. **Any TASK selected shall be evaluated in its entirety.**

NOTE: AREA OF OPERATION XI, Slow Flight, Stalls, and Spins, contains TASKs referred to as "proficiency" and "demonstration." The intent of TASKs A and B (proficiency) is to ensure that the flight instructor applicant is tested on proficiency for the purpose of teaching to students these TASKs that are required for pilot certification. The intent of TASKs C, D, E, G, and H (demonstration) is to ensure that the flight instructor applicant is knowledgeable and proficient in these maneuvers and can teach them to students for both familiarization and stall/spin awareness purposes.

With the exception of the **required** TASKs, the examiner shall not tell the applicant in advance, which TASKs will be included in the "plan of action." The applicant should be well prepared in **all** knowledge and skill areas included in the standards. Throughout the flight portion of the practical test, the examiner will evaluate the applicant's ability to simultaneously demonstrate and explain procedures and maneuvers, and to give flight instruction to students at various stages of flight training and levels of experience.

The purpose for including common errors in certain TASKs is to assist the examiner in determining that the flight instructor applicant has the ability to recognize, analyze, and correct such errors. The common errors listed in the TASK Objectives may or may not be found in the TASK References. However, the FAA considers their frequency of occurrence justification for their inclusion in the TASK Objectives.

The examiner is expected to use good judgment in the performance of simulated emergency procedures. The use of the safest means for simulation is expected. Consideration must be given to local conditions, both meteorological and topographical, at the time of the test, as well as the applicant's workload, and the condition of the aircraft used. If the procedure being evaluated would jeopardize safety, it is expected that the applicant will simulate that portion of the maneuver.

Special Emphasis Areas

Examiners shall place special emphasis upon areas of aircraft operation considered critical to flight safety. Among these are:

1. positive aircraft control;
2. positive exchange of the flight controls procedure;
3. stall/spin awareness;
4. collision avoidance;
5. wake turbulence avoidance;
6. LAHSO;
7. runway incursion avoidance;
8. CFIT;
9. ADM and risk management;
10. wire strike avoidance;
11. checklist usage;

12. temporary flight restrictions (TFRs);
13. special use airspace (SUA);
14. aviation security; and
15. other areas deemed appropriate to any phase of the practical test.

Although these areas may not be specifically addressed under each TASK, they are essential to flight safety and will be evaluated during the practical test. In all instances, the applicant's actions will relate to the complete situation.

Practical Test Prerequisites

An applicant for a flight instructor—initial certification practical test is required by 14 CFR part 61 to:

1. be at least 18 years of age;
2. be able to read, speak, write, and understand the English language. If there is a doubt, use AC 60-28, English Language Skill Standards;
3. hold either a commercial/instrument pilot or airline transport pilot certificate with an aircraft category rating appropriate to the flight instructor rating sought;
4. have an endorsement from an authorized instructor on the fundamentals of instructing appropriate to the required knowledge test;
5. have passed the appropriate flight instructor knowledge test(s) since the beginning of the 24th month before the month in which he or she takes the practical test. Knowledge test validity can be verified in FAA Order 8080.6, Conduct of Airman Knowledge Tests, Chapter 7, Eligibility Requirements; and
6. have an endorsement from an authorized instructor certifying that the applicant has been given flight training in the AREAS OF OPERATION listed in 14 CFR part 61, section 61.187 and a written statement from an authorized flight instructor within the preceding 60 days, in accordance with 14 CFR part 61, section 61.39, that instruction was given in preparation for the practical test. The endorsement shall also state that the instructor finds the applicant prepared for the required practical test, and that the applicant has demonstrated satisfactory knowledge of the subject area(s) in which the applicant was deficient on the airman knowledge test.

An applicant holding a flight instructor certificate who applies for an **additional** rating on that certificate is required by 14 CFR to:

1. hold a valid pilot certificate with ratings appropriate to the flight instructor rating sought;
2. have at least 15 hours as pilot-in-command in the category and class aircraft appropriate to the rating sought;
3. have passed the appropriate knowledge test prescribed for the issuance of a flight instructor certificate with the rating sought since the beginning of the 24th month before the month in which he/she takes the practical test; and
4. have an endorsement from an authorized instructor certifying that the applicant has been given flight training in the AREAS OF OPERATION listed in 14 CFR part 61, section 61.187 and a written statement from an authorized flight instructor within the preceding 60 days, in accordance with 14 CFR part 61, section 61.39, that instruction was given in preparation for the practical test. The endorsement shall also state that the instructor finds the applicant prepared for the required practical test, and that the applicant has demonstrated satisfactory knowledge of the subject area(s) in which the applicant was deficient on the airman knowledge test.

If there are questions concerning English language requirements, refer to your local FSDO or to AC 60-28, English Language Skill Standards Required by 14 CFR parts 61, 63, and 65. English language requirements should be determined to be met prior to beginning the practical test.

Aircraft and Equipment Required for the Practical Test

The flight instructor applicant is required by 14 CFR part 61, section 61.45 to provide an airworthy, certificated aircraft for use during the practical test. This section further requires that the aircraft must:

1. be of U.S., foreign or military registry of the same category, class, and type for the certificate and/or rating for which the applicant is applying;
2. have fully functioning dual controls except as provided in 14 CFR section 61.45(c) and (e); and
3. be capable of performing all appropriate TASKs for the flight instructor rating sought and have no operating limitations, which prohibit the performance of those TASKs. A complex airplane must be furnished for the performance of takeoff and landing maneuvers, and appropriate emergency procedures. A complex landplane is one having retractable gear, flaps, and controllable propeller. A complex seaplane is one having flaps and controllable propeller.

Flight Instructor Responsibility

An appropriately rated flight instructor is responsible for training the flight instructor applicant to acceptable standards in **all** subject matter areas, procedures, and maneuvers included in the TASKs within each AREA OF OPERATION in the appropriate flight instructor practical test standard.

Because of the impact of their teaching activities in developing safe, proficient pilots, flight instructors should exhibit a high level of knowledge, skill, and the ability to impart that knowledge and skill to students. The flight instructor must certify that the applicant is:

1. able to make a practical application of the fundamentals of instructing;
2. competent to teach the subject matter, procedures, and maneuvers included in the standards to students with varying backgrounds and levels of experience and ability;
3. able to perform the procedures and maneuvers included in the standards to at least the COMMERCIAL PILOT skill level while giving effective flight instruction; and
4. competent to pass the required practical test for the issuance of the flight instructor certificate with the associated category and class ratings or the addition of a category and/or class rating to a flight instructor certificate.

Throughout the applicant's training, the flight instructor is responsible for emphasizing the performance of, and the ability to teach, **effective visual scanning, runway incursion avoidance, collision avoidance procedures, and Land and Hold Short Operations (LAHSO)**. The flight instructor applicant should develop and use scenario based teaching methods particularly on special emphasis areas. These areas are covered in AC 90-48, Pilot's Role in Collision Avoidance; FAA-H-8083-3, Airplane Flying Handbook; FAA-H-8083-25, Pilot's Handbook of Aeronautical Knowledge; and the current Aeronautical Information Manual.

Examiner Responsibility

The examiner conducting the practical test is responsible for determining that the applicant meets acceptable standards of teaching ability, knowledge, and skill in the selected TASKs. The examiner makes this determination by accomplishing an Objective that is appropriate to each selected TASK, and includes an evaluation of the applicant's:

1. ability to apply the fundamentals of instructing;
2. knowledge of, and ability to teach, the subject matter, procedures, and maneuvers covered in the TASKs;
3. ability to perform the procedures and maneuvers included in the standards to the COMMERCIAL PILOT skill level while giving effective flight instruction; and
4. ability to analyze and correct common errors related to the procedures and maneuvers covered in the TASKs.

It is intended that oral questioning be used at any time during the ground or flight portion of the practical test to determine that the applicant can instruct effectively and has a comprehensive knowledge of the TASKs and their related safety factors.

During the flight portion of the practical test, the examiner shall act as a student during selected maneuvers. This will give the examiner an opportunity to evaluate the flight instructor applicant's ability to analyze and correct simulated common errors related to these maneuvers. The examiner will place special emphasis on the applicant's use of visual scanning and collision avoidance procedures, and the applicant's ability to teach those procedures.

Examiners should to the greatest extent possible test the applicant's application and correlation skills. When possible scenario based questions should be used during the practical test. The examiner will evaluate the applicant's ability to teach visual scanning, runway incursion avoidance, collision avoidance procedures, and Land and Hold Short Operations (LAHSO).

If the examiner determines that a TASK is incomplete, or the outcome uncertain, the examiner may require the applicant to repeat that TASK, or portions of that TASK. This provision has been made in the interest of fairness and does not mean that instruction, practice, or the repeating of an unsatisfactory task is permitted during the certification process. When practical, the remaining TASKs of the practical test phase should be completed before repeating the questionable TASK.

On multiengine practical tests where the failure of the most critical engine after lift off is required, the instructor applicant and examiner must give consideration to local atmospheric conditions, terrain and type of aircraft used. However the failure of an engine shall not be simulated until attaining at least Vsse/Vyse and at an altitude not lower than 200 feet AGL.

During simulated engine failures on multiengine practical tests, after simulated feathering of the propeller the engine shall be set to zero thrust. The examiner shall require the instructor applicant to simultaneously demonstrate and explain procedures for landing with a simulated feathered propeller with the engine set to zero thrust. **The examiner must not simulate any conditions that may jeopardize safe flight or result in possible damage to the aircraft**.

Satisfactory Performance

The practical test is passed if, in the judgment of the examiner, the applicant demonstrates satisfactory performance with regard to:

1. knowledge of the fundamentals of instructing;
2. knowledge of the technical subject areas;
3. knowledge of the flight instructor's responsibilities concerning the pilot certification process;
4. knowledge of the flight instructor's responsibilities concerning logbook entries and pilot certificate endorsements;
5. ability to demonstrate the procedures and maneuvers selected by the examiner to at least the COMMERCIAL PILOT skill level while giving effective instruction;
6. competence in teaching the procedures and maneuvers selected by the examiner;
7. competence in describing, recognizing, analyzing, and correcting common errors simulated by the examiner; and
8. knowledge of the development and effective use of a course of training, a syllabus, and a lesson plan.

Unsatisfactory Performance

If, in the judgment of the examiner, the applicant does not meet the standards of performance of any TASK performed, the applicable AREA OF OPERATION is considered unsatisfactory and therefore, the practical test is failed. The examiner or applicant may discontinue the test at any time when the failure of an AREA OF OPERATION makes the applicant ineligible for the certificate or rating sought. **The test will be continued only with the consent of the applicant.** If the test is discontinued, the applicant is entitled credit for only those AREAS OF OPERATION and their associated TASKs satisfactorily performed; however, during the retest and at the discretion of the examiner, any TASK may be re-evaluated, including those previously considered satisfactory. Specific reasons for disqualification are:

1. failure to perform a procedure or maneuver to the COMMERCIAL PILOT skill level while giving effective flight instruction;
2. failure to provide an effective instructional explanation while demonstrating a procedure or maneuver (explanation during the demonstration must be clear, concise, technically accurate, and complete with no prompting from the examiner);
3. any action or lack of action by the applicant which requires corrective intervention by the examiner to maintain safe flight;
4. failure to use proper and effective visual scanning techniques to clear the area before and while performing maneuvers.

When a Disapproval Notice is issued, the examiner must record the applicant's unsatisfactory performance in terms of AREA OF OPERATIONS and specific TASKS not meeting the standard appropriate to the practical test conducted. If the applicant fails the practical test because of a special emphasis area, the Notice of Disapproval shall indicate the associated TASK. An example would be: AREA OF OPERATION IX, Maneuvering During Slow Flight, failure to teach proper collision avoidance procedures.

Letter of Discontinuance

When a practical test is discontinued for reasons other than unsatisfactory performance (i.e., equipment failure, weather, or illness) FAA Form 8700-1, Airman Certificate and/or Rating Application, and, if applicable, the Airman Knowledge Test Report, is to be returned to the applicant. The examiner at that time prepares, signs, and issues a Letter of Discontinuance to the applicant. The Letter of Discontinuance should identify the AREAS OF OPERATION and their associated TASKs of the practical test that were successfully completed. The applicant should be advised that the Letter of Discontinuance must be presented to the examiner when the practical test is resumed, and made part of the certification file.

Aeronautical Decision Making And Risk Management

Throughout the practical test, the examiner evaluates the applicant's ability to use good aeronautical decision-making procedures in order to identify risks. The examiner accomplishes this requirement by developing scenarios that incorporate as many TASKs as possible to evaluate the applicants risk management in making safe aeronautical decisions. For example, the examiner may develop a scenario that incorporates weather decisions and performance planning.

The applicant's ability to utilize all the assets available in making a risk analysis to determine the safest course of action is essential for satisfactory performance. The scenarios should be realistic and within the capabilities of the aircraft used for the practical test.

Single-Pilot Resource Management

Single-Pilot Resource Management refers to the effective use of ALL available resources: human resources, hardware, and information. It is similar to Crew Resource Management (CRM) procedures that are being emphasized in multi-crewmember operations except that only one crewmember (the pilot) is involved. Human resources "...includes all other groups routinely working with the pilot who are involved in decisions that are required to operate a flight safely. These groups include, but are not limited to: dispatchers, weather briefers, maintenance personnel, and air traffic controllers." Pilot Resource Management is not a single TASK; it is a set of skill competencies that must be evident in all TASKs in this practical test standard as applied to single-pilot operation.

Applicant's Use of Checklists

Throughout the practical test, the instructor applicant is evaluated on the use and teaching of an appropriate checklist. Proper use is dependent on the specific TASK being evaluated. The situation may be such that the use of the checklist, while accomplishing elements of an Objective, would be either unsafe or impractical, especially in a single-pilot operation. In this case, a review of the checklist after the elements have been accomplished would be appropriate. Division of attention and proper visual scanning should be considered when using a checklist.

Use of Distractions During Practical Tests

Numerous studies indicate that many accidents have occurred when the pilot has been distracted during critical phases of flight. To evaluate the applicant's ability to utilize proper control technique while dividing attention both inside and outside the cockpit, the examiner shall cause realistic distractions using the flight portion of the practical test to evaluate the applicant's ability to divide attention while maintaining safe flight.

Positive Exchange of Flight Controls

During flight training, there must always be a clear understanding between students and flight instructors of who has control of the aircraft. Prior to flight, a briefing should be conducted that includes the procedure for the exchange of flight controls. A positive three-step process in the exchange of flight controls between pilots is a proven procedure and one that is strongly recommended.

When the instructor wishes the student to take control of the aircraft, he or she will say, "You have the flight controls." The student acknowledges immediately by saying, "I have the flight controls." The flight instructor again says, "You have the flight controls." When control is returned to the instructor, follow the same procedure. A visual check is recommended to verify that the exchange has occurred. There should never be any doubt as to who is flying the aircraft. The instructor applicant is expected to teach proper positive exchange of flight controls during the practical test.

Initial Flight Instructor Certification

An applicant who seeks initial flight instructor certification will be evaluated in all AREAS OF OPERATION of the standards appropriate to the rating(s) sought. The examiner shall refer to the **NOTE** in the front of the AREA OF OPERATION to determine which and how many TASKs shall be tested.

Addition of Aircraft Category and/or Class Ratings to a Flight Instructor Certificate

An applicant who holds a flight instructor certificate and seeks an additional aircraft category and/or class rating will be evaluated in at least the AREAS OF OPERATION and TASKs that are unique and appropriate to the rating(s) sought (see table at the beginning of each section). At the discretion of the examiner, the applicant's competence in **ALL** AREAS OF OPERATION may be evaluated.

Renewal or Reinstatement of a Flight Instructor Certificate

14 CFR part 61, sections 61.197(a)(1) and 61.199(a) allows an individual that holds a flight instructor certificate to renew or reinstate that certificate by passing a practical test. The examiner shall develop a plan of action that includes the AREAS OF OPERATION and at least the minimum number of TASKs prescribed in the table at the beginning of each section. The Renewal or Reinstatement of one rating on a Flight Instructor Certificate renews or reinstates all privileges existing on the certificate.

FLIGHT INSTRUCTOR

AIRPLANE—MULTIENGINE

Practical Test Standards

CONTENTS

Airplane Multiengine

CHECKLISTS

Applicant's Practical Test Checklist 2-v
Examiner's Checklist 2-vii

ADDITIONAL RATING TASK TABLE 2-xi

RENEWAL OR REINSTATEMENT OF A FLIGHT INSTRUCTOR TABLE 2-xii

AREAS OF OPERATION:

I. FUNDAMENTALS OF INSTRUCTING

A. The Learning Process 2-1
B. Human Behavior and Effective Communication 2-1
C. The Teaching Process 2-2
D. Teaching Methods 2-2
E. Critique and Evaluation 2-2
F. Flight Instructor Characteristics and Responsibilities 2-3
G. Planning Instructional Activity 2-3

II. TECHNICAL SUBJECT AREAS

A. Aeromedical Factors 2-4
B. Visual Scanning and Collision Avoidance 2-5
C. Principles of Flight 2-5
D. Airplane Flight Controls 2-5
E. Airplane Weight and Balance 2-6
F. Navigation and Flight Planning 2-6
G. Night Operations 2-7
H. High Altitude Operations 2-7
I. Federal Aviation Regulations and Publications 2-8
J. National Airspace System 2-8
K. Navigation Aids and Radar Services 2-9
L. Logbook Entries and Certificate Endorsements 2-9
M. Water and Seaplane Characteristics 2-10
N. Seaplane Bases, Rules, and Aids to Marine Navigation 2-10

III. PREFLIGHT PREPARATION

A. Certificates and Documents 2-11
B. Weather Information 2-11
C. Airworthiness Requirements 2-12

IV. PREFLIGHT LESSON ON A MANEUVER TO BE PERFORMED IN FLIGHT

A. Maneuver Lesson 2-13

V. PREFLIGHT PROCEDURES

A. Preflight Inspection 2-14
B. Single-Pilot Crew Resource Management 2-15
C. Engine Starting 2-16
D. Taxiing—Landplane 2-17
E. Taxiing—Seaplane 2-18
F. Sailing 2-19
G. Before Takeoff Check 2-20

VI. AIRPORT AND SEAPLANE BASE OPERATIONS

A. Radio Communications and ATC Light Signals 2-21
B. Traffic Patterns 2-22
C. Airport/Seaplane Base, Runway and Taxiway Signs, Markings, and Lighting 2-23

VII. TAKEOFFS, LANDINGS, AND GO-AROUNDS

A. Normal and Crosswind Takeoff and Climb 2-24
B. Short-Field (Confined Area AMES) Takeoff and Maximum Performance and Climb 2-25
C. Glassy-Water Takeoff and Climb 2-26
D. Rough-Water Takeoff and Climb 2-27
E. Normal and Crosswind Approach and Landing 2-28
F. Go-Around/Rejected Landing 2-29
G. Short-Field (Confined Area AMES) Approach and Landing 2-30
H. Glassy-Water Approach and Landing 2-31
I. Rough-Water Approach and Landing 2-32

VIII. **FUNDAMENTALS OF FLIGHT**

A. Straight-and-Level Flight.......... 2-33
B. Level Turns.......... 2-34
C. Straight Climbs and Climbing Turns.......... 2-35
D. Straight Descents and Descending Turns.......... 2-36

IX. **PERFORMANCE MANEUVERS**

A. Steep Turns.......... 2-37

X. **GROUND REFERENCE MANEUVERS**

A. Rectangular Course.......... 2-38
B. S-Turns Across a Road.......... 2-39
C. Turns Around a Point.......... 2-40

XI. **SLOW FLIGHT AND STALLS**

A. Maneuvering During Slow Flight.......... 2-41
B. Power-On Stalls.......... 2-42
C. Power-Off Stalls.......... 2-43
D. Accelerated Maneuver Stalls (Demonstration).......... 2-44

XII. **BASIC INSTRUMENT MANEUVERS**

A. Straight-and-Level Flight.......... 2-45
B. Constant Airspeed Climbs.......... 2-46
C. Constant Airspeed Descents.......... 2-47
D. Turns to Headings.......... 2-48
E. Recovery from Unusual Flight Attitudes.......... 2-49

XIII. **EMERGENCY OPERATIONS**

A. Systems and Equipment Malfunctions.......... 2-50
B. Engine Failure During Takeoff Before V_{MC}.......... 2-51
C. Engine Failure After Lift-Off.......... 2-52
D. Approach and Landing with an Inoperative Engine.......... 2-53
E. Emergency Descent.......... 2-54
F. Emergency Equipment and Survival Gear.......... 2-55

XIV. MULTIENGINE OPERATIONS

A. Operation of Systems 2-56
B. Performance and Limitations 2-56
C. Flight Principles—Engine Inoperative 2-57
D. Maneuvering with One Engine Inoperative 2-58
E. Vmc Demonstration 2-60
F. Demonstrating the Effects of Various Airspeeds and Configurations During Engine Inoperative Performance 2-61

XV. POSTFLIGHT PROCEDURES

A. Postflight Procedures 2-62
B. Anchoring 2-63
C. Docking and Mooring 2-64
D. Beaching 2-65
E. Ramping 2-66

APPLICANT'S PRACTICAL TEST CHECKLIST

APPOINTMENT WITH INSPECTOR OR EXAMINER:

NAME: ____

TIME/DATE: ____

ACCEPTABLE AIRCRAFT:

- ❑ View-Limiting Device (if applicable)
- ❑ Aircraft Documents:
  - Airworthiness Certificate
  - Registration Certificate
  - Operating Limitations
- ❑ Aircraft Maintenance Records:
  - Airworthiness Inspections
- ❑ Pilot's Operating Handbook, FAA-Approved Airplane Flight Manual

PERSONAL EQUIPMENT

- ❑ Current Aeronautical Charts
- ❑ Computer and Plotter
- ❑ Flight Plan Form
- ❑ Flight Logs
- ❑ Current AIM
- ❑ Current Airport Facility Directory

PERSONAL RECORDS

- ❑ Pilot Certificate
- ❑ Medical Certificate
- ❑ Completed FAA Form 8710-1, Airman Certificate and/or Rating Application
- ❑ Airman Knowledge Test Report
- ❑ Logbook with Instructor's Endorsement
- ❑ Letter of Discontinuance (if applicable)
- ❑ Notice of Disapproval (if applicable)
- ❑ Approved School Graduation Certificate (if applicable)
- ❑ Examiner's Fee (if applicable)

EXAMINER'S CHECKLIST
FLIGHT INSTRUCTOR—AIRPLANE (MULTIENGINE)

APPLICANT'S NAME ______________________________

EXAMINER'S NAME ______________________________

DATE __________ **TYPE CHECK** ______________________

TYPE AIRPLANE ________________________

AREA OF OPERATION:

I. FUNDAMENTALS OF INSTRUCTING

- ❑ **A.** The Learning Process
- ❑ **B.** Human Behavior and Effective Communication
- ❑ **C.** The Teaching Process
- ❑ **D.** Teaching Methods
- ❑ **E.** Critique and Evaluation
- ❑ **F.** Flight Instructor Characteristics and Responsibilities
- ❑ **G.** Planning Instructional Activity

II. TECHNICAL SUBJECT AREAS

- ❑ **A.** Aeromedical Factors
- ❑ **B.** Visual Scanning and Collision Avoidance
- ❑ **C.** Principles of Flight
- ❑ **D.** Airplane Flight Controls
- ❑ **E.** Airplane Weight and Balance
- ❑ **F.** Navigation and Flight Planning
- ❑ **G.** Night Operations
- ❑ **H.** High Altitude Operations
- ❑ **I.** Federal Aviation Regulations and Publications
- ❑ **J.** National Airspace System
- ❑ **K.** Navigation Aids and Radar Services
- ❑ **L.** Logbook Entries and Certificate Endorsements
- ❑ **M.** Water and Seaplane Characteristics
- ❑ **N.** Seaplane Bases, Rules, and Aids to Marine Navigation

III. PREFLIGHT PREPARATION

- ❑ **A.** Certificates and Documents
- ❑ **B.** Weather Information
- ❑ **C.** Airworthiness Requirements

IV. PREFLIGHT LESSON ON A MANEUVER TO BE PERFORMED IN FLIGHT

- ❑ Maneuver Lesson

V. PREFLIGHT PROCEDURES

- ❑ **A.** Preflight Inspection
- ❑ **B.** Single-Pilot Resource Management
- ❑ **C.** Engine Starting
- ❑ **D.** Taxiing—Landplane
- ❑ **E.** Taxiing—Seaplane
- ❑ **F.** Sailing
- ❑ **G.** Before Takeoff Check

VI. AIRPORT AND SEAPLANE BASE OPERATIONS

- ❑ **A.** Radio Communications and ATC Light Signals
- ❑ **B.** Traffic Patterns
- ❑ **C.** Airport/Seaplane Base, Runway and Taxiway, Signs, Markings, and Lighting

VII. TAKEOFFS, LANDINGS AND, GOAROUNDS

- ❑ **A.** Normal and Crosswind Takeoff and Climb
- ❑ **B.** Short-Field (Confined Area (AMES)) Takeoff and Maximum Performance Climb
- ❑ **C.** Glassy-Water Takeoff and Climb
- ❑ **D.** Rough-Water Takeoff and Climb
- ❑ **E.** Normal and Crosswind Approach and Landing
- ❑ **F.** Go-Around/Rejected Landing
- ❑ **G.** Short-Field (Confined Area (AMES)) Approach and Landing
- ❑ **H.** Glassy-Water Approach and Landing
- ❑ **I.** Rough-Water Approach and Landing

VIII. FUNDAMENTALS OF FLIGHT

- ❑ **A.** Straight-and-Level Flight
- ❑ **B.** Level Turns
- ❑ **C.** Straight Climbs and Climbing Turns
- ❑ **D.** Straight Descents and Descending Turns

IX. PERFORMANCE MANEUVERS

- ❑ Steep Turns

X. GROUND REFERENCE MANEUVERS

- ❑ **A.** Rectangular Course
- ❑ **B.** S-Turns Across a Road
- ❑ **C.** Turns Around a Point

XI. SLOW FLIGHT AND STALLS

- ❑ **A.** Maneuvering During Slow Flight
- ❑ **B.** Power-On Stalls
- ❑ **C.** Power-Off Stalls
- ❑ **D.** Accelerated Maneuver Stalls (Demonstration)

XII. BASIC INSTRUMENT MANEUVERS

- ❑ **A.** Straight-and-Level Flight
- ❑ **B.** Constant Airspeed Climbs
- ❑ **C.** Constant Airspeed Descents
- ❑ **D.** Turns to Headings
- ❑ **E.** Recovery from Unusual Flight Attitudes

XIII. EMERGENCY OPERATIONS

- ❑ **A.** Systems and Equipment Malfunctions
- ❑ **B.** Engine Failure During Takeoff Before V_{MC}
- ❑ **C.** Engine Failure After Lift-Off
- ❑ **D.** Approach and Landing with an Inoperative Engine
- ❑ **E.** Emergency Descent
- ❑ **F.** Emergency Equipment and Survival Gear

XIV. MULTIENGINE OPERATIONS

- ❑ **A.** Operations of Systems
- ❑ **B.** Performance and Limitations
- ❑ **C.** Flight Principles—Engine Inoperative
- ❑ **D.** Maneuvering with One Engine Inoperative
- ❑ **E.** V_{MC} Demonstration
- ❑ **F.** Demonstrating the Effects of Various Airspeeds and Configurations During Engine Inoperative Performance

XV. POSTFLIGHT PROCEDURES

- ❑ **A.** Postflight Procedures
- ❑ **B.** Anchoring
- ❑ **C.** Docking and Mooring
- ❑ **D.** Beaching
- ❑ **E.** Ramping

ADDITIONAL RATING TASK TABLE

ADDITION OF A MULTIENGINE CLASS RATING (AND AN AIRPLANE CATEGORY RATING, IF APPROPRIATE) TO A FLIGHT INSTRUCTOR CERTIFICATE						
REQUIRED AREAS OF OPERATION	FLIGHT INSTRUCTOR CERTIFICATE AND RATING HELD					
	ASE	RH	RG	G	IA	IH
I	NONE	NONE	NONE	NONE	NONE	NONE
II	NONE	C,D	C,D	C,D	C,D	C,D
III	NONE	NONE	NONE	NONE	NONE	NONE
IV	NONE	NONE	NONE	NONE	NONE	NONE
V	*	*	*	*	*	*
VI	NONE	*	NONE	*	*	*
VII	*	*	*	*	*	*
VIII	NONE	*	*	*	*	*
IX	*	*	*	*	*	*
X	NONE	*	*	*	*	*
XI	*	*	*	*	*	*
XII	NONE	*	*	*	NONE	*
XIII	*	*	*	*	*	*
XIV	*	*	*	*	*	*
XV	NONE	*	*	*	*	*

NOTE: If an applicant holds more than one rating on a flight instructor certificate and the table indicates both a "NONE" and a "SELECT ONE" for a particular AREA OF OPERATION, the "NONE" entry applies. This is logical since the applicant has satisfactorily accomplished the AREA OF OPERATION on a previous flight instructor practical test. At the discretion of the examiner, the applicant's competence in **any** AREAS OF OPERATION may be evaluated.

* Refer to NOTE under Area of Operation for Task requirements.

RENEWAL OR REINSTATEMENT OF A FLIGHT INSTRUCTOR TABLE

Airplane Multiengine Category

REQUIRED AREAS OF OPERATION	NUMBER OF TASKS
II	TASK L and 1other TASK
III	1
IV	1
VII	2 Takeoffs and 2 Landings
IX	1
X	1
XI	1
XIII	2
XIV	1

The Renewal or Reinstatement of one rating on a Flight Instructor Certificate renews or reinstates all privileges existing on the certificate. (14 CFR part 61, sections 61.197 and 61.199)

I. AREA OF OPERATION: FUNDAMENTALS OF INSTRUCTING

NOTE: The examiner shall select TASK F and one other TASK.

A. TASK: THE LEARNING PROCESS

REFERENCE: FAA-H-8083-9.

Objective. To determine that the applicant exhibits instructional knowledge of the elements of the learning process by describing:

1. Learning theory.
2. Characteristics of learning.
3. Principles of learning.
4. Levels of learning.
5. Learning physical skills.
6. Memory.
7. Transfer of learning.

B. TASK: HUMAN BEHAVIOR AND EFFECTIVE COMMUNICATION

REFERENCE: FAA-H-8083-9.

Objective. To determine that the applicant exhibits instructional knowledge of the elements of the teaching process by describing:

1. Human behavior—

    a. control of human behavior.
    b. human needs.
    c. defense mechanisms.
    d. the flight instructor as a practical psychologist.

2. Effective communication—

    a. basic elements of communication.
    b. barriers of effective communication.
    c. developing communication skills.

C. TASK: THE TEACHING PROCESS

REFERENCE: FAA-H-8083-9.

Objective. To determine that the applicant exhibits instructional knowledge of the elements of the teaching process by describing:

1. Preparation of a lesson for a ground or flight instructional period.
2. Presentation methods.
3. Application, by the student, of the knowledge and skills presented.
4. Review and evaluation of student performance.

D. TASK: TEACHING METHODS

REFERENCE: FAA-H-8083-9.

Objective. To determine that the applicant exhibits instructional knowledge of the elements of teaching methods by describing:

1. Material organization.
2. The lecture method.
3. The cooperative or group learning method.
4. The guided discussion method.
5. The demonstration-performance method.
6. Computer-based training method.

E. TASK: CRITIQUE AND EVALUATION

REFERENCE: FAA-H-8083-9.

Objective. To determine that the applicant exhibits instructional knowledge of the elements of critique and evaluation by explaining:

1. Critique—

    a. purpose and characteristics of an effective critique.
    b. methods and ground rules for a critique.

2. Evaluation—

    a. characteristics of effective oral questions and what types to avoid.
    b. responses to student questions.
    c. characteristics and development of effective written questions.
    d. characteristics and uses of performance tests, specifically, the FAA practical test standards.

F. TASK: FLIGHT INSTRUCTOR CHARACTERISTICS AND RESPONSIBILITIES

REFERENCE: FAA-H-8083-9.

Objective. To determine that the applicant exhibits instructional knowledge of the elements of flight instructor characteristics and responsibilities by describing:

1. Aviation instructor responsibilities in—
   a. providing adequate instruction.
   b. establishing standards of performance.
   c. emphasizing the positive.
2. Flight instructor responsibilities in—
   a. providing student pilot evaluation and supervision.
   b. preparing practical test recommendations and endorsements.
   c. determining requirements for conducting additional training and endorsement requirements.
3. Professionalism as an instructor by—
   a. explaining important personal characteristics.
   b. describing methods to minimize student frustration.

G. TASK: PLANNING INSTRUCTIONAL ACTIVITY

REFERENCE: FAA-H-8083-9.

Objective. To determine that the applicant exhibits instructional knowledge of the elements of planning instructional activity by describing:

1. Developing objectives and standards for a course of training.
2. Theory of building blocks of learning.
3. Requirements for developing a training syllabus.
4. Purpose and characteristics of a lesson plan.

II. AREA OF OPERATION: TECHNICAL SUBJECT AREAS

NOTE: The examiner shall select TASK L and at least one other TASK.

A. TASK: AEROMEDICAL FACTORS

REFERENCES: FAA-H-8083-3; FAA-S-8081-12, FAA-S-8081-14; AIM.

Objective. To determine that the applicant exhibits instructional knowledge of the elements related to aeromedical factors by describing:

1. How to obtain an appropriate medical certificate.
2. How to obtain a medical certificate in the event of a possible medical deficiency.
3. The causes, symptoms, effects, and corrective action of the following medical factors—

    a. hypoxia.
    b. hyperventilation.
    c. middle ear and sinus problems.
    d. spatial disorientation.
    e. motion sickness.
    f. carbon monoxide poisoning.
    g. fatigue and stress.
    h. dehydration.

4. The effects of alcohol and drugs, and their relationship to flight safety.
5. The effect of nitrogen excesses during scuba dives and how this affects pilots and passengers during flight.

B. TASK: VISUAL SCANNING AND COLLISION AVOIDANCE

REFERENCES: FAA-H-8083-25, AC 90-48, FAA-H-8083-3; AIM.

Objective. To determine that the applicant exhibits instructional knowledge of the elements of visual scanning and collision avoidance by describing:

1. Relationship between a pilot's physical condition and vision.
2. Environmental conditions that degrade vision.
3. Vestibular and visual illusions.
4. "See and avoid" concept.
5. Proper visual scanning procedure.
6. Relationship between poor visual scanning habits and increased collision risk.
7. Proper clearing procedures.
8. Importance of knowing aircraft blind spots.
9. Relationship between aircraft speed differential and collision risk.
10. Situations which involve the greatest collision risk.

C. TASK: PRINCIPLES OF FLIGHT

REFERENCES: FAA-H-8083-3, FAA-H-8083-25.

Objective. To determine that the applicant exhibits instructional knowledge of the elements of principles of flight by describing:

1. Airfoil design characteristics.
2. Airplane stability and controllability.
3. Turning tendency (torque effect).
4. Load factors in airplane design.
5. Wingtip vortices and precautions to be taken.

D. TASK: AIRPLANE FLIGHT CONTROLS

REFERENCES: FAA-H-8083-3, FAA-H-8083-25.

Objective. To determine that the applicant exhibits instructional knowledge of the elements related to the airplane flight controls by describing the purpose, location, direction of movement, effect and proper procedure for use of the:

1. Primary flight controls.
2. Trim control/s.
3. Wing flaps.

E. TASK: AIRPLANE WEIGHT AND BALANCE

REFERENCES: FAA-H-8083-1, FAA-H-8083-3, FAA-H-8083-25; FAA-S-8081-12, FAA-S-8081-14.

Objective. To determine that the applicant exhibits instructional knowledge of the elements of airplane weight and balance by describing:

1. Weight and balance terms.
2. Effect of weight and balance on performance.
3. Methods of weight and balance control.
4. Determination of total weight and center of gravity and the changes that occur when adding, removing, or shifting weight.

F. TASK: NAVIGATION AND FLIGHT PLANNING

REFERENCES: FAA-H-8083-3, FAA-H-8083-25; FAA-S-8081-12, FAA-S-8081-14.

Objective. To determine that the applicant exhibits instructional knowledge of the elements of navigation and flight planning by describing:

1. Terms used in navigation.
2. Features of aeronautical charts.
3. Importance of using the proper and current aeronautical charts.
4. Method of plotting a course, selection of fuel stops and alternates, and appropriate actions in the event of unforeseen situations.
5. Fundamentals of pilotage and dead reckoning.
6. Fundamentals of radio navigation.
7. Diversion to an alternate.
8. Lost procedures.
9. Computation of fuel consumption.
10. Importance of preparing and properly using a flight log.
11. Importance of a weather check and the use of good judgment in making a "go/no-go" decision.
12. Purpose of, and procedure used in, filing a flight plan.
13. Global positioning system (GPS) navigation.

G. TASK: NIGHT OPERATIONS

REFERENCES: FAA-H-8083-3, FAA-H-8083-25; FAA-S-8081-12, FAA-S-8081-14; AIM.

Objective. To determine that the applicant exhibits instructional knowledge of the elements of night operations by describing:

1. Factors related to night vision.
2. Disorientation and night optical illusions.
3. Proper adjustment of interior lights.
4. Importance of having a flashlight with a red lens.
5. Night preflight inspection.
6. Engine starting procedures, including use of position and anticollision lights prior to start.
7. Taxiing and orientation on an airport.
8. Takeoff and climb-out.
9. In-flight orientation.
10. Importance of verifying the airplane's attitude by reference to flight instruments.
11. Night emergencies procedures.
12. Traffic patterns.
13. Approaches and landings with and without landing lights.
14. Go-arounds.

H. TASK: HIGH ALTITUDE OPERATIONS

REFERENCES: 14 CFR part 91; AC 61-107, FAA-H-8083-3; FAA-S-8081-12; POH/AFM, AIM.

Objective. To determine that the applicant exhibits instructional knowledge of the elements of high altitude operations by describing:

1. Regulatory requirements for use of oxygen.
2. Physiological hazards associated with high altitude operations.
3. Characteristics of a pressurized airplane and various types of supplemental oxygen systems.
4. Importance of "aviators breathing oxygen."
5. Care and storage of high-pressure oxygen bottles.
6. Problems associated with rapid decompression and corresponding solutions.

I. TASK: FEDERAL AVIATION REGULATIONS AND PUBLICATIONS

REFERENCES: 14 CFR parts 1, 61, 91; NTSB part 830; FAA-H-8083-25; POH/AFM, AIM.

Objective. To determine that the applicant exhibits instructional knowledge of the elements related to Federal Aviation Regulations publications by describing:

1. Availability and method of revision of 14 CFR parts 1, 61, 91, and NTSB part 830—

    a. purpose.
    b. general content.

2. Availability of flight information publications, advisory circulars, practical test standards, pilot operating handbooks, and FAA-approved airplane flight manuals by describing—

    a. availability.
    b. purpose.
    c. general content.

J. TASK: NATIONAL AIRSPACE SYSTEM

REFERENCES: 14 CFR part 91; FAA-S-8081-12, FAA-S-8081-14; AIM.

Objective. To determine that the applicant exhibits instructional knowledge of the elements of the national airspace system by describing:

1. Basic VFR weather minimums—for all classes of airspace.
2. Airspace classes—their operating rules, pilot certification and airplane equipment requirements for the following—

    a. Class A.
    b. Class B.
    c. Class C.
    d. Class D.
    e. Class E.
    f. Class G.

3. Special use airspace (SUA).
4. Temporary flight restrictions (TFR).

K. TASK: NAVIGATION AIDS AND RADAR SERVICES

REFERENCES: FAA-H-8083-3, FAA-H-8083-15; FAA-S-8081-12, FAA-S-8081-14; AIM.

Objective: To determine that the applicant exhibits instructional knowledge of the elements related to navigation aids and radar service by describing:

1. One ground-based navigational aid (VOR/VORTAC, NDB, DME, LORAN).
2. Satellite-based navigation aids.
3. Radar service and procedures.
4. Global positioning system (GPS).

L. TASK: LOGBOOK ENTRIES AND CERTIFICATE ENDORSEMENTS

REFERENCES: 14 CFR part 61; AC 61-65.

Objective. To determine that the applicant exhibits instructional knowledge of the elements related to logbook entries and certificate endorsements by describing:

1. Required logbook entries for instruction given.
2. Required student pilot certificate endorsements, including appropriate logbook entries.
3. Preparation of a recommendation for a pilot practical test, including appropriate logbook entry for—

    a. initial pilot certification.
    b. additional pilot certification
    c. additional aircraft qualification.

4. Required endorsement of a pilot logbook for the satisfactory completion of the required FAA flight review.
5. Required flight instructor records.

M. TASK: WATER AND SEAPLANE CHARACTERISTICS (AMES)

REFERENCES: FAA-H-8083-3, FAA-H-8083-23; FAA-S-8081-12, FAA-S-8081-14; Seaplane Manual.

Objective. To determine that the applicant exhibits instructional knowledge of the elements related to water and seaplane characteristics by describing:

1. The characteristics of water surface as affected by features, such as—

    a. size and location of water operating area.
    b. protected and unprotected operating areas.
    c. surface wind.
    d. direction and height of waves.
    e. direction and strength of water current.
    f. floating and partially submerged debris.
    g. sandbars, islands, and shoals.

2. Seaplane float or hull construction and its relationship to performance.
3. Causes of porpoising and skipping and pilot action necessary to prevent or to correct those occurrences.

N. TASK: SEAPLANE BASES, RULES, AND AIDS TO MARINE NAVIGATION (AMES)

REFERENCES: 14 CFR part 91; AC 61-69, FAA-H-8083-3, FAA-H-8083-23; FAA-S-8081-12, FAA-S-8081-14; USCG Navigation Rules, International -Inland.

Objective. To determine that the applicant exhibits instructional knowledge of the elements related to seaplane bases, rules, and aids to marine navigation by describing:

1. How to locate and identify seaplane bases on charts or in directories.
2. Operating restrictions at various seaplane bases.
3. Right-of-way, steering, and sailing rules pertinent to seaplane operation.
4. Purpose and identification of marine navigation aids such as buoys, beacons, lights, and sound signals.

III. AREA OF OPERATION: PREFLIGHT PREPARATION

NOTE: The examiner shall select at least one TASK.

A. TASK: CERTIFICATES AND DOCUMENTS

REFERENCES: 14 CFR parts 23, 43, 61, 91; FAA-H-8083-3, FAA-H-8083-25; FAA-S-8081-12, FAA-S-8081-14; POH/AFM.

Objective. To determine that the applicant exhibits instructional knowledge of the elements related to certificates and documents by describing:

1. The training requirements for the issuance of a recreational, private, and commercial pilot certificate.
2. The privileges and limitations of pilot certificates and ratings at recreational, private, and commercial levels.
3. Class and duration of medical certificates.
4. Recent pilot flight experience requirements.
5. Required entries in pilot logbook or flight record.

B. TASK: WEATHER INFORMATION

REFERENCES: AC 00-6, AC 00-45, FAA-H-8083-25; FAA-S-8081-12, FAA-S-8081-14.

Objective. To determine that the applicant exhibits instructional knowledge of the elements related to weather information by describing:

1. Importance of a thorough preflight weather briefing.
2. Means and sources of obtaining weather information.
3. Use of real time weather reports, forecasts, and charts for developing scenario based training.
4. Inflight weather advisories.
5. Recognition of aviation weather hazards to include wind shear.
6. Factors to be considered in making a "go/no-go" decision.

C. TASK: AIRWORTHINESS REQUIREMENTS

REFERENCES: 14 CFR parts 39, 43; FAA-S-8081-12, FAA-S-8081-14; POH/AFM.

Objective. To determine that the applicant exhibits instructional knowledge of the elements related to required airworthiness by explaining:

1. Required instruments and equipment for day/night VFR.
2. Procedures and limitations for determining airworthiness of the airplane with inoperative instruments and equipment without minimum equipment list (MEL).
3. Requirements and procedures for obtaining a special flight permit.
4. Airworthiness directives, compliance records, maintenance/inspection requirements, and appropriate records.
5. Procedures for deferring maintenance on aircraft without an approved MEL.

IV. AREA OF OPERATION: PREFLIGHT LESSON ON A MANEUVER TO BE PERFORMED IN FLIGHT

NOTE: Examiner shall select at least one maneuver TASK from AREAS OF OPERATION VII through XIV, and ask the applicant to present a preflight lesson on the selected maneuver as the lesson would be taught to a student.

A. TASK: MANEUVER LESSON

REFERENCES: FAA-H-8083-3, FAA-H-8083-9, FAA-H-8083-23, FAA-H-8083-25; FAA-S-8081-12, FAA-S-8081-14; POH/AFM.

Objective. To determine that the applicant exhibits instructional knowledge of the selected maneuver by:

1. Stating the purpose.
2. Giving an accurate, comprehensive oral description, including the elements and common errors.
3. Using instructional aids, as appropriate.
4. Describing the recognition, analysis, and correction of common errors.

V. AREA OF OPERATION: PREFLIGHT PROCEDURES

NOTE: The examiner shall select at least one TASK.

A. TASK: PREFLIGHT INSPECTION (AMEL and AMES)

REFERENCES: AC 61-84; FAA-H-8083-3, FAA-H-8083-23; FAA-S-8081-12, FAA-S-8081-14; POH/AFM.

Objective. To determine that the applicant:

1. Exhibits instructional knowledge of the elements of a preflight inspection, as applicable to the airplane used for the practical test, by describing—
   a. reasons for the preflight inspection, items that should be inspected, and how defects are detected.
   b. importance of using the appropriate checklist.
   c. how to determine fuel and oil quantity and contamination.
   d. detection of fuel, oil, and hydraulic leaks.
   e. inspection of the oxygen system, including supply and proper operation (if applicable).
   f. inspection of the flight controls and water rudder (if applicable).
   g. detection of visible structural damage.
   h. removal of tie-downs, control locks, and wheel chocks.
   i. removal of ice and frost.
   j. importance of the proper loading and securing of baggage, cargo, and equipment.
   k. use of sound judgment in determining whether the airplane is in an airworthy condition for safe flight.
2. Exhibits instructional knowledge of common errors related to a preflight inspection by describing—
   a. failure to use or the improper use of a checklist.
   b. hazards which may result from allowing distractions to interrupt a visual inspection.
   c. inability to recognize discrepancies to determine airworthiness.
   d. failure to ensure servicing with the proper fuel and oil.
   e. failure to ensure proper loading and securing of baggage, cargo, and equipment.
3. Demonstrates and simultaneously explains a preflight inspection from an instructional standpoint.

B. TASK: SINGLE-PILOT CREW RESOURCE MANAGEMENT (AMEL and AMES)

REFERENCES: FAA-H-8083-3; FAA-S-8081-12, FAA-S-8081-14; POH/AFM.

Objective. To determine that the applicant:

1. Exhibits instructional knowledge of the elements of single-pilot crew resource management by describing—

   a. proper arranging and securing of essential materials and equipment in the cockpit.
   b. proper use and/or adjustment of cockpit items such as safety belts, shoulder harnesses, rudder pedals, and seats.
   c. occupant briefing on emergency procedures and use of safety belts.
   d. detection of fuel, oil, and hydraulic leaks.

2. Exhibits instructional knowledge of common errors related to single-pilot crew resource management by describing—

   a. failure to place and secure essential materials and equipment for easy access during flight.
   b. failure to properly use and/or adjustment of cockpit items, such as safety belts, shoulder harnesses, rudder pedals, and seats.
   c. failure to provide improper adjustment of equipment and controls. occupant briefing on emergency procedures and use of safety belts.
   d. failure to provide occupant briefing on emergency procedures and use of safety belts.
   e. failure to ensure proper loading and securing of baggage, cargo, and equipment.

3. Demonstrates and simultaneously explains single-pilot crew resource management from an instructional standpoint.

C. TASK: ENGINE STARTING (AMEL and AMES)

REFERENCES: FAA-H-8083-3, FAA-H-8083-23, FAA-H-8083-25; AC 91-13, AC 91-55; FAA-S-8081-12, FAA-S-8081-14; POH/AFM.

Objective. To determine that the applicant:

1. Exhibits instructional knowledge of the elements of engine starting, as appropriate to the airplane used for the practical test by describing—

   a. safety precautions related to starting.
   b. use of external power.
   c. effect of atmospheric conditions on starting.
   d. importance of following the appropriate checklist.
   e. adjustment of engine controls during start.
   f. prevention of airplane movement during and after start.

2. Exhibits instructional knowledge of common errors related to engine starting by describing—

   a. failure to properly use the appropriate checklist.
   b. failure to use safety precautions related to starting.
   c. improper adjustment of engine controls during start.
   d. failure to ensure proper clearance of the propeller.

3. Demonstrates and simultaneously explains engine starting from an instructional standpoint.

D. TASK: TAXIING—LANDPLANE (AMEL)

REFERENCES: FAA-H-8083-3; FAA-S-8081-12, FAA-S-8081-14; POH/AFM.

Objective. To determine that the applicant:

1. Exhibits instructional knowledge of the elements of landplane taxiing by describing—

   a. proper brake check and correct use of brakes.
   b. compliance with airport/taxiway surface marking, signals, and ATC clearances or instructions.
   c. how to control direction and speed.
   d. flight control positioning for various wind conditions.
   e. procedures used to avoid other aircraft and hazards.
   f. avoiding runway incursions.

2. Exhibits instructional knowledge of common errors related to landplane taxiing by describing—

   a. improper use of brakes.
   b. improper positioning of the flight controls for various wind conditions.
   c. hazards of taxiing too fast.
   d. failure to comply with airport/taxiway surface marking, signals, and ATC clearances or instructions.

3. Demonstrates and simultaneously explains landplane taxiing from an instructional standpoint.
4. Analyzes and corrects simulated common errors related to landplane taxiing.

E. TASK: TAXIING—SEAPLANE (AMES)

REFERENCES: FAA-H-8083-3, FAA-H-8083-23; FAA-S-8081-12; FAA-S-8081-14; POH/AFM; USCG Navigation Rules, International - Inland.

Objective. To determine that the applicant:

1. Exhibits instructional knowledge of the elements of seaplane taxiing by describing—
   a. wind effect.
   b. prevention of porpoising and skipping.
   c. selection of the most suitable course for taxiing, following available marking aids.
   d. conditions where idle, plowing, and step taxiing are used.
   e. procedures for idle, plowing, and step taxiing.
   f. control positioning for various wind conditions.
   g. use of water rudders.
   h. procedures used to avoid other aircraft and hazards.
   i. procedures used to avoid excessive water spray into the propeller.
   j. 180° and 360° turns in idle, plowing, and step positions.
   k. application of right-of-way rules.
2. Exhibits instructional knowledge of common errors related to seaplane taxiing by describing—
   a. improper positioning of flight controls for various wind conditions.
   b. improper control of speed and direction.
   c. failure to prevent porpoising and skipping.
   d. failure to use the most suitable course and available marking aids.
   e. failure to use proper clearing procedures to avoid hazards.
   f. failure to apply right-of-way rules.
3. Demonstrates and simultaneously explains seaplane taxiing from an instructional standpoint.
4. Analyzes and corrects simulated common errors related to seaplane taxiing.

F. TASK: SAILING (AMES)

REFERENCES: FAA-H-8083-3, FAA-H-8083-23; FAA-S-8081-12, FAA-S-8081-14; POH/AFM; USCG Navigation Rules, International - Inland.

Objective. To determine that the applicant:

1. Exhibits instructional knowledge of the elements of sailing by describing—
    a. procedures used in sailing (engine idling or shut down, as appropriate).
    b. conditions and situations where sailing would be used.
    c. selection of the most favorable course to follow.
    d. use of flight controls, flaps, doors, and water rudders to follow the selected course.
    e. procedures used to change direction from downwind to crosswind.
    f. control of speed.
2. Exhibits instructional knowledge of common errors related to sailing by describing—
    a. failure to select the most favorable course to destination.
    b. improper use of controls, flaps, and water rudders.
    c. improper procedure when changing direction.
    d. improper procedure for speed control.
3. Demonstrates and simultaneously explains sailing from an instructional standpoint.
4. Analyzes and corrects simulated common errors related to sailing.

G. TASK: BEFORE TAKEOFF CHECK (AMEL and AMES)

REFERENCES: FAA-H-8083-3, FAA-H-8083-23; FAA-S-8081-12, FAA-S-8081-14; POH/AFM.

Objective. To determine that the applicant:

1. Exhibits instructional knowledge of the elements of the before takeoff check by describing—
   a. positioning the airplane with consideration for other aircraft, surface conditions, and wind.
   b. division of attention inside and outside the cockpit.
   c. importance of following the checklist and responding to each checklist item.
   d. reasons for ensuring suitable engine temperatures and pressures for run-up and takeoff.
   e. method used to determine the airplane is in a safe operating condition.
   f. importance of reviewing takeoff performance airspeeds, expected takeoff distances, and emergency procedures.
   g. method used for ensuring that the takeoff area or path is free of hazards.
   h. method of avoiding runway incursions and ensuring no conflict with traffic prior to taxiing into takeoff position.
2. Exhibits instructional knowledge of common errors related to the before takeoff check by describing—
   a. failure to properly use the appropriate checklist.
   b. improper positioning of the airplane.
   c. improper acceptance of marginal engine performance.
   d. an improper check of flight controls.
   e. hazards of failure to review takeoff and emergency procedures.
   f. failure to avoid runway incursions and to ensure no conflict with traffic prior to taxiing into takeoff position.
3. Demonstrates and simultaneously explains a before takeoff check from an instructional standpoint.
4. Analyzes and corrects simulated common errors related to a before takeoff check.

VI. AREA OF OPERATION: AIRPORT AND SEAPLANE BASE OPERATIONS

NOTE: The examiner shall select at least one TASK.

A. TASK: RADIO COMMUNICATIONS AND ATC LIGHT SIGNALS
(AMEL and AMES)

REFERENCES: FAA-H-8083-25, FAA-H-8083-3; FAA-S-8081-12, FAA-S-8081-14; AIM.

Objective. To determine that the applicant:

1. Exhibits instructional knowledge of the elements of radio communications and ATC light signals by describing—
   a. selection and use of appropriate radio frequencies.
   b. recommended procedure and phraseology for radio communications.
   c. receipt, acknowledgement of, and compliance with, ATC clearances and instructions.
   d. interpretation of, and compliance with, ATC light signals.
2. Exhibits instructional knowledge of common errors related to radio communications and ATC light signals by describing—
   a. use of improper frequencies.
   b. improper procedure and phraseology when using radio communications.
   c. failure to acknowledge, or properly comply with, ATC clearances and instructions.
   d. failure to understand, or to properly comply with, ATC light signals.
3. Demonstrates and simultaneously explains radio communication procedures from an instructional standpoint.
4. Analyzes and corrects simulated common errors related to radio communications and ATC light signals.

B. TASK: TRAFFIC PATTERNS (AMEL and AMES)

REFERENCES: FAA-H-8083-25, FAA-H-8083-3, AC 90-42, AC 90-66; FAA-S-8081-12, FAA-S-8081-14; AIM.

Objective. To determine that the applicant:

1. Exhibits instructional knowledge of the elements of traffic patterns by describing—
   a. operations at airports and seaplane bases with and without operating control towers.
   b. adherence to traffic pattern procedures, instructions, and rules.
   c. how to maintain proper spacing from other traffic.
   d. how to maintain the desired ground track.
   e. wind shear and wake turbulence avoidance procedures.
   f. orientation with the runway or landing area in use.
   g. how to establish a final approach at an appropriate distance from the runway or landing area.
   h. use of checklist.
2. Exhibits instructional knowledge of common errors related to traffic patterns by describing—
   a. failure to comply with traffic pattern instructions, procedures, and rules.
   b. improper correction for wind drift.
   c. inadequate spacing from other traffic.
   d. poor altitude or airspeed control.
3. Demonstrates and simultaneously explains traffic patterns from an instructional standpoint.
4. Analyzes and corrects simulated common errors related to traffic patterns.

C. TASK: AIRPORT/SEAPLANE BASE, RUNWAY AND TAXIWAY SIGNS, MARKINGS, AND LIGHTING (AMEL and AMES)

REFERENCES: FAA-H-8083-23, FAA-H-8083-25; FAA-S-8081-12, FAA-S-8081-14; AIM; AC 150/5340-1, AC 150/5340-18.

Objective. To determine that the applicant exhibits instructional knowledge of the elements of airport/seaplane base, runway and taxiway signs, markings, and lighting by describing:

1. Exhibits instructional knowledge of the elements of airport/seaplane base, runway and taxiway signs, markings, and lighting by describing—

    a. identification and proper interpretation of airport/seaplane base, runway and taxiway signs and markings with emphasis on runway incursion avoidance.
    b. identification and proper interpretation of airport/seaplane base, runway and taxiway lighting with emphasis on runway incursion avoidance.

2. Exhibits instructional knowledge of common errors related to airport/seaplane base, runway and taxiway signs, markings, and lighting by describing—

    a. failure to comply with airport/seaplane base, runway and taxiway signs and markings.
    b. failure to comply with airport/seaplane base, runway and taxiway lighting.
    c. failure to use proper runway incursion avoidance procedures.

3. Demonstrates and simultaneously explains airport/seaplane base, runway and taxiway signs, markings, and lighting from an instructional standpoint.
4. Analyzes and corrects simulated common errors related to explains airport/seaplane base, runway and taxiway signs, markings, and lighting.

VII. AREA OF OPERATION: TAKEOFFS, LANDINGS, AND GO-AROUNDS

NOTE: The examiner shall select at least two takeoffs and two landings TASKs.

A. TASK: NORMAL AND CROSSWIND TAKEOFF AND CLIMB (AMEL and AMES)

REFERENCES: FAA-H-8083-3, FAA-H-8083-23; FAA-S-8081-12, FAA-S-8081-14; POH/AFM.

Objective. To determine that the applicant:

1. Exhibits instructional knowledge of the elements of a normal and crosswind takeoff and climb by describing—

    a. procedures before taxiing onto the runway or takeoff area to ensure runway incursion avoidance.
    b. normal and crosswind takeoff procedures.
    c. difference between a normal and a glassy-water takeoff (seaplane).
    d. normal and crosswind lift-off procedures.
    e. proper climb attitude, power setting, and airspeed (V_Y).
    f. proper use of checklist.

2. Exhibits instructional knowledge of common errors related to a normal and crosswind takeoff and climb by describing—

    a. improper runway incursion avoidance procedures.
    b. improper use of controls during a normal or crosswind takeoff.
    c. inappropriate lift-off procedures.
    d. improper climb attitude, power setting and airspeed (V_Y).
    e. improper use of checklist.

3. Demonstrates and simultaneously explains a normal or a crosswind takeoff and climb from an instructional standpoint.
4. Analyzes and corrects simulated common errors related to a normal or a crosswind takeoff and climb.

B. TASK: SHORT-FIELD (CONFINED AREA (AMES)) TAKEOFF AND MAXIMUM PERFORMANCE AND CLIMB (AMEL and AMES)

REFERENCES: FAA-H-8083-3, FAA-H-8083-23; FAA-S-8081-12, FAA-S-8081-14; POH/AFM.

Objective. To determine that the applicant:

1. Exhibits instructional knowledge of the elements of a short-field takeoff and climb by describing—

    a. procedures before taxiing onto the runway or takeoff area to ensure runway incursion avoidance.
    b. short-field takeoff procedures.
    c. short-field lift-off procedures.
    d. initial climb attitude and airspeed (V_X) until obstacle is cleared (50 feet AGL) .
    e. proper use of checklist.

2. Exhibits instructional knowledge of common errors related to a short-field takeoff and climb by describing—

    a. improper runway incursion avoidance procedures.
    b. improper use of controls during a short-field takeoff.
    c. improper lift-off procedures.
    d. improper initial climb attitude, power setting and airspeed (V_X) to clear obstacle.
    e. improper use of checklist.

3. Demonstrates and simultaneously explains a short-field takeoff and climb from an instructional standpoint.
4. Analyzes and corrects simulated common errors related to a short-field takeoff and climb.

C. TASK: GLASSY-WATER TAKEOFF AND CLIMB (ASES)

REFERENCES: FAA-H-8083-3, FAA-H-8083-23; FAA-S-8081-12, FAA-S-8081-14; POH/AFM.

Objective. To determine that the applicant:

1. Exhibits instructional knowledge of the elements of a glassy-water takeoff and climb by describing—

    a. procedures before taxiing onto the takeoff area to ensure waterway is clear of objects or obstructions.
    b. flight control, flap and water rudder use during glassy-water takeoff procedures.
    c. appropriate planning attitude and lift-off procedures on glassy-water.
    d. initial climb attitude and airspeed (V_X), if an obstacle is present (50 feet AGL) or (V_Y).
    e. proper use of after takeoff checklist.

2. Exhibits instructional knowledge of common errors related to a glassy-water takeoff and climb by describing—

    a. improper takeoff water way clearance procedures.
    b. poor judgment in the selection of a suitable takeoff area.
    c. improper use of controls during a glassy-water takeoff.
    d. inappropriate lift-off procedures.
    e. hazards of inadvertent contact with the water after becoming airborne.
    f. improper climb attitude, power setting, and airspeed (V_Y) or (V_X).
    g. improper use of checklist.

3. Demonstrates and simultaneously explains a glassy-water takeoff and climb from an instructional standpoint.
4. Analyzes and corrects simulated common errors related to a glassy-water takeoff and climb.

D. TASK: ROUGH-WATER TAKEOFF AND CLIMB (AMES)

REFERENCES: FAA-H-8083-3, FAA-H-8083-23; FAA-S-8081-12, FAA-S-8081-14; POH/AFM.

Objective. To determine that the applicant:

1. Exhibits instructional knowledge of the elements of a rough-water takeoff and climb by describing—

    a. procedures before taxiing onto the takeoff area to ensure waterway is clear of objects or obstructions.
    b. flight control, flap and water rudder use during rough-water takeoff procedures.
    c. appropriate planning attitude and lift-off procedures on rough-water.
    d. initial climb attitude and airspeed (V_X).if an obstacle is present (50 feet AGL) or (V_Y).
    e. proper use of after takeoff checklist.

2. Exhibits instructional knowledge of common errors related to a rough-water takeoff and climb by describing—

    a. improper takeoff water way clearance procedures.
    b. poor judgment in the selection of a suitable takeoff area.
    c. improper use of controls during a rough-water takeoff.
    d. inappropriate lift-off procedures.
    e. hazards of inadvertent contact with the water after becoming airborne.
    f. improper climb attitude, power setting, and airspeed (V_Y) or (V_X).
    g. improper use of checklist.

3. Demonstrates and simultaneously explains a rough-water takeoff and climb from an instructional standpoint.
4. Analyzes and corrects simulated common errors related to a rough-water takeoff and climb.

E. TASK: NORMAL AND CROSSWIND APPROACH AND LANDING (AMEL and AMES)

REFERENCES: FAA-H-8083-3, FAA-H-8083-23; FAA-S-8081-12, FAA-S-8081-14; POH/AFM.

Objective. To determine that the applicant:

1. Exhibits instructional knowledge of the elements of a normal and a crosswind approach and landing by describing—

   a. how to determine landing performance and limitations.
   b. configuration, power, and trim.
   c. obstructions and other hazards, which should be considered.
   d. a stabilized approach at the recommended airspeed to the selected touchdown area.
   e. course of action if selected touchdown area is going to be missed.
   f. coordination of flight controls.
   g. a precise ground track.
   h. wind shear and wake turbulence.
   i. crosswind procedure.
   j. timing, judgment, and control procedure during roundout and touchdown.
   k. directional control after touchdown.
   l. use of brakes (landplane).
   m. use of checklist.

2. Exhibits instructional knowledge of common errors related to a normal and a crosswind approach and landing by describing—

   a. improper use of landing performance data and limitations.
   b. failure to establish approach and landing configuration at appropriate time or in proper sequence.
   c. failure to establish and maintain a stabilized approach.
   d. inappropriate removal of hand from throttles.
   e. improper procedure during roundout and touchdown.
   f. poor directional control after touchdown.
   g. improper use of brakes (landplane).

3. Demonstrates and simultaneously explains a normal or a crosswind approach and landing from an instructional standpoint.
4. Analyzes and corrects simulated common errors related to a normal or crosswind approach and landing.

F. TASK: GO-AROUND/REJECTED LANDING (AMEL and AMES)

REFERENCES: FAA-H-8083-3, FAA-H-8083-23; FAA-S-8081-12, FAA-S-8081-14; POH/AFM.

Objective. To determine that the applicant:

1. Exhibits instructional knowledge of the elements of a go-around/rejected landing by describing—

    a. situations where a go-around is necessary.
    b. importance of making a prompt decision.
    c. importance of applying takeoff power immediately after the go-around decision is made.
    d. importance of establishing proper pitch attitude.
    e. wing flaps retraction.
    f. use of trim.
    g. landing gear retraction.
    h. proper climb speed.
    i. proper track and obstruction clearance.
    j. use of checklist.
    k. importance of manufacturer's recommended procedures.

2. Exhibits instructional knowledge of common errors related to a go-around/rejected landing by describing—

    a. failure to recognize a situation where a go-around/rejected landing is necessary.
    b. hazards of delaying a decision to go around.
    c. improper power application.
    d. failure to control pitch attitude.
    e. failure to compensate for torque effect.
    f. improper trim technique.
    g. failure to maintain recommended airspeeds.
    h. improper wing flaps or landing gear retraction procedure.
    i. failure to maintain proper track during climb-out.
    j. failure to remain well clear of obstructions and other traffic.

3. Demonstrates and simultaneously explains a go-around/rejected landing from an instructional standpoint.
4. Analyzes and corrects simulated common errors related to a go-around/rejected landing.

G. TASK: SHORT-FIELD (CONFINED AREA—AMES) APPROACH AND LANDING (AMEL and AMES)

REFERENCES: FAA-H-8083-3, FAA-H-8083-23; FAA-S-8081-12, FAA-S-8081-14; POH/AFM.

Objective. To determine that the applicant:

1. Exhibits instructional knowledge of the elements of a short-field approach and landing by describing—
    a. how to determine landing performance and limitations.
    b. configuration and trim.
    c. proper use of pitch and power to maintain desired approach angle.
    d. barriers and other hazards which should be considered.
    e. effect of wind.
    f. selection of touchdown and go-around points.
    g. a stabilized approach at the recommended airspeed to the selected touchdown point.
    h. coordination of flight controls.
    i. a precise ground track.
    j. timing, judgment, and control procedure during roundout and touchdown.
    k. directional control after touchdown.
    l. use of brakes. (Landplane)
    m. use of checklist.
2. Exhibits instructional knowledge of common errors related to a short-field approach and landing by describing—
    a. improper use of landing performance data and limitations.
    b. failure to establish approach and landing configuration at appropriate time or in proper sequence.
    c. failure to establish and maintain a stabilized approach.
    d. improper technique in use of power, wing flaps, and trim.
    e. inappropriate removal of hand from throttles.
    f. improper procedure during roundout and touchdown.
    g. poor directional control after touchdown.
    h. improper use of brakes. (Landplane)
3. Demonstrates and simultaneously explains a short-field approach and landing from an instructional standpoint.
4. Analyzes and corrects simulated common errors related to a short-field approach and landing.

H. TASK: GLASSY-WATER APPROACH AND LANDING (AMES)

REFERENCES: FAA-H-8083-3, FAA-H-8083-23; FAA-S-8081-12, FAA-S-8081-14; POH/AFM.

Objective. To determine that the applicant:

1. Exhibits instructional knowledge of the elements of a glassy-water approach and landing by describing—

   a. how to determine landing performance and limitations.
   b. configuration and trim.
   c. deceptive characteristics of glassy water.
   d. selection of a suitable landing area and go-around point.
   e. terrain and obstructions which should be considered.
   f. detection of hazards in the landing area such as shallow water, obstructions, or floating debris.
   g. coordination of flight controls.
   h. a precise ground track.
   i. a power setting and pitch attitude that will result in the recommended airspeed and rate of descent throughout the final approach to touchdown.
   j. how to maintain positive after-landing control.
   k. use of checklist.

2. Exhibits instructional knowledge of common errors related to a glassy-water approach and landing by describing—

   a. improper use of landing performance data and limitations.
   b. failure to establish approach and landing configuration at appropriate time and in proper sequence.
   c. failure to establish and maintain a stabilized approach at the recommended airspeed and rate of descent.
   d. improper technique in use of power, wing flaps, or trim.
   e. inappropriate removal of hand from throttles.
   f. failure to touch down with power in the proper stabilized attitude.
   g. failure to maintain positive after-landing control.

3. Demonstrates and simultaneously explains a glassy-water approach and landing from an instructional standpoint.
4. Analyzes and corrects simulated common errors related to a glassy-water approach and landing.

I. TASK: ROUGH-WATER APPROACH AND LANDING (AMES)

REFERENCES: FAA-H-8083-3, FAA-H-8083-23; FAA-S-8081-12, FAA-S-8081-14; POH/AFM.

Objective. To determine that the applicant:

1. Exhibits instructional knowledge of the elements of a rough-water approach and landing by describing—
   a. how to determine landing performance and limitations.
   b. review of wind conditions.
   c. how landing area characteristics can be evaluated.
   d. selection of a suitable landing area and go-around point.
   e. terrain and obstructions which should be considered.
   f. detection of hazards in the landing area such as shallow water, obstructions, or floating debris.
   g. configuration and trim.
   h. coordination of flight controls.
   i. a precise ground track.
   j. a stabilized approach at the recommended airspeed to the selected touchdown area.
   k. timing, judgment, and control procedure during roundout and touchdown.
   l. maintenance of positive after-landing control.
   m. use of checklist.
2. Exhibits instructional knowledge of common errors related to a rough-water approach and landing by describing—
   a. improper use of landing performance data and limitations.
   b. failure to establish approach and landing configuration at appropriate time and in proper sequence.
   c. failure to establish and maintain a stabilized approach.
   d. improper procedure in use of power, wing flaps, or trim.
   e. inappropriate removal of hand from throttles.
   f. improper procedure during roundout and touchdown.
   g. failure to maintain positive after-landing control.
3. Demonstrates and simultaneously explains a rough-water approach and landing from an instructional standpoint.
4. Analyzes and corrects simulated common errors related a rough-water approach and landing.

VIII. AREA OF OPERATION: FUNDAMENTALS OF FLIGHT

NOTE: The examiner shall select at least one TASK.

A. TASK: STRAIGHT-AND-LEVEL FLIGHT (AMEL and AMES)

REFERENCES: FAA-H-8083-3, FAA-H-8083-23; FAA-S-8081-14.

Objective. To determine that the applicant:

1. Exhibits instructional knowledge of the elements of straight-and-level flight by describing—

    a. effect and use of flight controls.
    b. the Integrated Flight Instruction method.
    c. outside and instrument references used for pitch, bank, and power control; the crosscheck and interpretation of those references; and the control technique used.
    d. trim technique.
    e. methods that can be used to overcome tenseness and overcontrolling.

2. Exhibits instructional knowledge of common errors related to straight-and-level flight by describing—

    a. failure to crosscheck and correctly interpret outside and instrument references.
    b. application of control movements rather than pressures.
    c. uncoordinated use of flight controls.
    d. faulty trim technique.

3. Demonstrates and simultaneously explains straight-and-level flight from an instructional standpoint.
4. Analyzes and corrects simulated common errors related to straight-and-level flight.

B. TASK: LEVEL TURNS (AMEL and AMES)

REFERENCES: FAA-H-8083-3; FAA-S-8081-14.

Objective. To determine that the applicant:

1. Exhibits instructional knowledge of the elements of level turns by describing—
   a. effect and use of flight controls.
   b. the Integrated Flight Instruction method.
   c. outside and instrument references used for pitch, bank, and power control; the crosscheck and interpretation of those references; and the control technique used.
   d. trim technique.
   e. methods that can be used to overcome tenseness and overcontrolling.
2. Exhibits instructional knowledge of common errors related to level turns by describing—
   a. failure to crosscheck and correctly interpret outside and instrument references.
   b. application of control movements rather than pressures.
   c. uncoordinated use of flight controls.
   d. faulty altitude and bank control.
3. Demonstrates and simultaneously explains a level turn from an instructional standpoint.
4. Analyzes and corrects simulated common errors related to level turns.

C. TASK: STRAIGHT CLIMBS AND CLIMBING TURNS (AMEL and AMES)

REFERENCES: FAA-H-8083-3; FAA-S-8081-14.

Objective. To determine that the applicant:

1. Exhibits instructional knowledge of the elements of straight climbs and climbing turns by describing—
   a. effect and use of flight controls.
   b. the Integrated Flight Instruction method.
   c. outside and instrument references used for pitch, bank, and power control; the crosscheck and interpretation of those references; and the control technique used.
   d. trim technique.
   e. methods that can be used to overcome tenseness and overcontrolling.
2. Exhibits instructional knowledge of common errors related to straight climbs and climbing turns by describing—
   a. failure to crosscheck and correctly interpret outside and instrument references.
   b. application of control movements rather than pressures.
   c. improper correction for torque effect.
   d. faulty trim technique.
3. Demonstrates and simultaneously explains straight climbs and climbing turns from an instructional standpoint.
4. Analyzes and corrects simulated common errors related to straight climbs and climbing turns.

D. TASK: STRAIGHT DESCENTS AND DESCENDING TURNS (AMEL and AMES)

REFERENCES: FAA-H-8083-3; FAA-S-8081-14.

Objective. To determine that the applicant:

1. Exhibits instructional knowledge of the elements of straight descents and descending turns by describing—
   a. effect and use of flight controls.
   b. the Integrated Flight Instruction method.
   c. outside and instrument references used for pitch, bank, and power control; the crosscheck and interpretation of those references; and the control technique used.
   d. trim technique.
   e. methods that can be used to overcome tenseness and overcontrolling.
2. Exhibits instructional knowledge of common errors related to straight descents and descending turns by describing—
   a. failure to crosscheck and correctly interpret outside and instrument references.
   b. application of control movements rather than pressures.
   c. uncoordinated use of flight controls.
   d. faulty trim technique.
   e. failure to clear engine and use carburetor heat, as appropriate.
3. Demonstrates and simultaneously explains straight descents and descending turns from an instructional standpoint.
4. Analyzes and corrects simulated common errors related to straight descents and descending turns.

IX. AREA OF OPERATION: PERFORMANCE MANEUVERS

A. TASK: STEEP TURNS (AMEL and AMES)

REFERENCES: FAA-H-8083-3; FAA-S-8081-12, FAA-S-8081-14; POH/AFM.

Objective. To determine that the applicant:

1. Exhibits instructional knowledge of the elements of steep turns by describing—

    a. relationship of bank angle, load factor, and stalling speed.
    b. over-banking tendency.
    c. torque effect in right and left turns.
    d. selection of a suitable altitude.
    e. orientation, division of attention, and planning.
    f. entry and rollout procedure.
    g. coordination of flight and power controls.
    h. altitude, bank, and power control during the turn.
    i. proper recovery to straight-and-level flight.

2. Exhibits instructional knowledge of common errors related to steep turns by describing—

    a. improper pitch, bank, and power coordination during entry and rollout.
    b. uncoordinated use of flight controls.
    c. improper procedure in correcting altitude deviations.
    d. loss of orientation.

3. Demonstrates and simultaneously explains steep turns from an instructional standpoint.
4. Analyzes and corrects simulated common errors related to steep turns.

X. AREA OF OPERATION: GROUND REFERENCE MANEUVERS

NOTE: The examiner shall select at least one TASK.

A. TASK: RECTANGULAR COURSE (AMEL and AMES)

REFERENCES: FAA-H-8083-3; FAA-S-8081-14.

Objective. To determine that the applicant:

1. Exhibits instructional knowledge of the elements of a rectangular course by describing—

    a. how to select a safe altitude.
    b. how to select a suitable ground reference with consideration given to emergency landing areas.
    c. orientation, division of attention, and planning.
    d. configuration and airspeed prior to entry.
    e. relationship of a rectangular course to an airport traffic pattern.
    f. wind drift correction.
    g. how to maintain desired altitude, airspeed, and distance from ground reference boundaries.
    h. timing of turn entries and rollouts.
    i. coordination of flight controls.

2. Exhibits instructional knowledge of common errors related to a rectangular course by describing—

    a. poor planning, orientation, or division of attention.
    b. uncoordinated flight control application.
    c. improper correction for wind drift.
    d. failure to maintain selected altitude or airspeed.
    e. selection of a ground reference where there is no suitable emergency landing area within gliding distance.

3. Demonstrates and simultaneously explains a rectangular course from an instructional standpoint.
4. Analyzes and corrects simulated common errors related to a rectangular course.

B. TASK: S-TURNS ACROSS A ROAD (AMEL and AMES)

REFERENCES: FAA-H-8083-3; FAA-S-8081-14.

Objective. To determine that the applicant:

1. Exhibits instructional knowledge of the elements of S-turns across a road by describing—
   a. how to select a safe altitude.
   b. how to select a suitable ground reference line with consideration given to emergency landing areas.
   c. orientation, division of attention, and planning.
   d. configuration and airspeed prior to entry.
   e. entry procedure.
   f. wind drift correction.
   g. tracking of semicircles of equal radii on either side of the selected ground reference line.
   h. how to maintain desired altitude and airspeed.
   i. turn reversal over the ground reference line.
   j. coordination of flight controls.
2. Exhibits instructional knowledge of common errors related to S-turns across a road by describing—
   a. faulty entry technique.
   b. poor planning, orientation, or division of attention.
   c. uncoordinated flight control application.
   d. improper correction for wind drift.
   e. an unsymmetrical ground track.
   f. failure to maintain selected altitude or airspeed.
   g. selection of a ground reference line where there is no suitable emergency landing area within gliding distance.
3. Demonstrates and simultaneously explains S-turns across a road from an instructional standpoint.
4. Analyzes and corrects simulated common errors related to S-turns across a road.

C. TASK: TURNS AROUND A POINT (AMEL and AMES)

REFERENCES: FAA-H-8083-3; FAA-S-8081-14.

Objective. To determine that the applicant:

1. Exhibits instructional knowledge of the elements of turns around a point by describing—
   a. how to select a safe altitude.
   b. how to select a suitable ground reference point with consideration given to emergency landing areas.
   c. orientation, division of attention, and planning.
   d. configuration and airspeed prior to entry.
   e. entry procedure.
   f. wind drift correction.
   g. how to maintain desired altitude, airspeed, and distance from reference point.
   h. coordination of flight controls.
2. Exhibits instructional knowledge of common errors related to turns around a point by describing—
   a. faulty entry procedure.
   b. poor planning, orientation, or division of attention.
   c. uncoordinated flight control application.
   d. improper correction for wind drift.
   e. failure to maintain selected altitude or airspeed.
   f. selection of a ground reference point where there is no suitable emergency landing area within gliding distance.
3. Demonstrates and simultaneously explains turns around a point from an instructional standpoint.
4. Analyzes and corrects simulated common errors related to turns around a point.

XI. AREA OF OPERATION: SLOW FLIGHT AND STALLS

NOTE: The examiner shall select at least one TASK. Stalls shall not be performed with one engine at reduced power or inoperative and the other engine(s) developing effective power.

Stalls using high power settings should not be performed. The high pitch angles necessary to induce these stalls could possibly result in uncontrollable flight.

Examiners and instructors should be alert to the possible development of high sink rates when performing stalls in multiengine airplanes with high wing loading.

A. TASK: MANEUVERING DURING SLOW FLIGHT (AMEL and AMES)

REFERENCES: FAA-H-8083-3; FAA-S-8081-12, FAA-S-8081-14; POH/AFM.

Objective. To determine that the applicant:

1. Exhibits instructional knowledge of the elements of maneuvering during slow flight by describing—

    a. relationship of configuration, weight, center of gravity, maneuvering loads, angle of bank, and power to flight characteristics and controllability.
    b. relationship of the maneuver to critical flight situations, such as go-arounds.
    c. performance of the maneuver with selected landing gear and flap configurations in straight-and-level flight and level turns.
    d. specified airspeed for the maneuver.
    e. coordination of flight controls.
    f. trim technique.
    g. re-establishment of cruise flight.

2. Exhibits instructional knowledge of common errors related to maneuvering during slow flight by describing—

    a. failure to establish specified gear and flap configuration.
    b. improper entry technique.
    c. failure to establish and maintain the specified airspeed.
    d. excessive variations of altitude and heading when a constant altitude and heading are specified.
    e. rough or uncoordinated control technique.
    f. improper correction for torque effect.
    g. improper trim technique.
    h. unintentional stalls.
    i. inappropriate removal of hand from throttles.

3. Demonstrates and simultaneously explains maneuvering during slow flight from an instructional standpoint.
4. Analyzes and corrects simulated common errors related to maneuvering during slow flight.

B. TASK: POWER-ON STALLS (AMEL and AMES)

REFERENCES: FAA-H-8083-3; FAA-S-8081-12, FAA-S-8081-14; POH/AFM.

Objective. To determine that the applicant:

1. Exhibits instructional knowledge of the elements of power-on stalls, in climbing flight (straight or turning), with selected landing gear and flap configurations by describing—

   a. aerodynamics of power-on stalls.
   b. relationship of various factors, such as landing gear and flap configuration, weight, center of gravity, load factor, and bank angle to stall speed.
   c. flight situations where unintentional power-on stalls may occur.
   d. entry technique and minimum entry altitude.
   e. performance of power-on stalls in climbing flight (straight or turning).
   f. coordination of flight controls.
   g. recognition of the first indications of power-on stalls.
   h. recovery technique and minimum recovery altitude.

2. Exhibits instructional knowledge of common errors related to power-on stalls, in climbing flight (straight or turning), with selected landing gear and flap configurations by describing—

   a. failure to establish the specified landing gear and flap configuration prior to entry.
   b. improper pitch, heading, and bank control during straight-ahead stalls.
   c. improper pitch and bank control during turning stalls.
   d. rough or uncoordinated control technique.
   e. failure to recognize the first indications of a stall.
   f. failure to achieve a stall.
   g. improper torque correction.
   h. poor stall recognition and delayed recovery.
   i. excessive altitude loss or excessive airspeed during recovery.
   j. secondary stall during recovery.

3. Demonstrates and simultaneously explains power-on stalls, in climbing flight (straight or turning), with selected landing gear and flap configurations, from an instructional standpoint.
4. Analyzes and corrects simulated common errors related to power-on stalls, in climbing flight (straight and turning), with selected landing gear and flap configurations.

C. TASK: POWER-OFF STALLS (AMEL and AMES)

REFERENCES: FAA-H-8083-3; FAA-S-8081-12, FAA-S-8081-14; POH/AFM.

Objective. To determine that the applicant:

1. Exhibits instructional knowledge of the elements of power-off stalls, in descending flight (straight or turning), with selected landing gear and flap configurations by describing—

    a. aerodynamics of power-off stalls.
    b. relationship of various factors, such as landing gear and flap configuration, weight, center of gravity, load factor, and bank angle to stall speed.
    c. flight situations where unintentional power-off stalls may occur.
    d. entry technique and minimum entry altitude.
    e. performance of power-off stalls in descending flight (straight or turning).
    f. coordination of flight controls.
    g. recognition of the first indications of power-off stalls.
    h. recovery technique and minimum recovery altitude.

2. Exhibits instructional knowledge of common errors related to power-off stalls, in descending flight (straight or turning), with selected landing gear and flap configurations by describing—

    a. failure to establish the specified landing gear and flap configuration prior to entry.
    b. improper pitch, heading, and bank control during straight-ahead stalls.
    c. improper pitch and bank control during turning stalls.
    d. rough or uncoordinated control technique.
    e. failure to recognize the first indications of a stall.
    f. failure to achieve a stall.
    g. improper torque correction.
    h. poor stall recognition and delayed recovery.
    i. excessive altitude loss or excessive airspeed during recovery.
    j. secondary stall during recovery.

3. Demonstrates and simultaneously explains power-off stalls, in descending flight (straight or turning), with selected landing gear and flap configurations, from an instructional standpoint.
4. Analyzes and corrects simulated common errors related to power-off stalls, in descending flight (straight or turning), with selected landing gear and flap configurations.

D. TASK: ACCELERATED MANEUVER STALLS (DEMONSTRATION) (AMEL and AMES)

NOTE: This TASK shall be completed by oral examination or demonstrated at examiner's discretion.

REFERENCES: FAA-H-8083-3; POH/AFM.

Objective. To determine that the applicant:

1. Exhibits instructional knowledge of the elements of accelerated maneuver stalls by describing—

   a. aerodynamics of accelerated maneuver stalls.
   b. flight situations where accelerated maneuver stalls may occur.
   c. hazards of accelerated stalls during stall or spin recovery.
   d. entry procedure and minimum entry altitude.
   e. recognition of the accelerated stall.
   f. recovery procedure and minimum recovery altitude.

2. Demonstrates and simultaneously explains accelerated maneuver stall, from an instructional standpoint—
3. Exhibits instructional knowledge of common errors related to accelerated maneuver stalls by describing—

   a. failure to establish proper configuration prior to entry.
   b. improper or inadequate demonstration of the recognition of and recovery from an accelerated maneuver stall.
   c. Failure to present simulated student instruction that adequately emphasizes the hazards of poor procedures in recovering from an accelerated stall.

4. Analyzes and corrects simulated common errors related to accelerated stalls.

XII. AREA OF OPERATION: BASIC INSTRUMENT MANEUVERS

NOTE: The examiner shall select at least one TASK.

A. TASK: STRAIGHT-AND-LEVEL FLIGHT (AMEL and AMES)

REFERENCES: FAA-H-8083-3, FAA-H-8083-15; FAA-S-8081-14.

Objective. To determine that the applicant:

1. Exhibits instructional knowledge of the elements of straight-and-level flight, solely by reference to instruments by describing—

    a. instrument crosscheck, instrument interpretation, and aircraft control.
    b. instruments used for pitch, bank, and power control, and how those instruments are used to maintain altitude, heading, and airspeed.
    c. trim technique.

2. Exhibits instructional knowledge of common errors related to straight-and-level flight solely by reference to instruments by describing—

    a. "fixation," "omission," and "emphasis" errors during instrument crosscheck.
    b. improper instrument interpretation.
    c. improper control applications.
    d. failure to establish proper pitch, bank, or power adjustments during altitude, heading, or airspeed corrections.
    e. faulty trim technique.

3. Demonstrates and simultaneously explains straight-and-level flight, solely by reference to instruments, from an instructional standpoint.
4. Analyzes and corrects simulated common errors related to straight-and-level flight, solely by reference to instruments.

B. TASK: **CONSTANT AIRSPEED CLIMBS** (AMEL and AMES)

REFERENCES: FAA-H-8083-3, FAA-H-8083-15; FAA-S-8081-14.

Objective. To determine that the applicant:

1. Exhibits instructional knowledge of the elements of straight and turning, constant airspeed climbs, solely by reference to instruments by describing—

    a. instrument crosscheck, instrument interpretation, and aircraft control.
    b. instruments used for pitch, bank, and power control during entry, during the climb, and during level off, and how those instruments are used to maintain climb heading and airspeed.
    c. trim technique.

2. Exhibits instructional knowledge of common errors related to straight and turning, constant airspeed climbs, solely by reference to instruments by describing—

    a. "fixation," "omission," and "emphasis" errors during instrument crosscheck.
    b. improper instrument interpretation.
    c. improper control applications.
    d. failure to establish proper pitch, bank, or power adjustments during heading and airspeed corrections.
    e. improper entry or level-off technique.
    f. faulty trim technique.

3. Demonstrates and simultaneously explains a straight, constant airspeed climb, solely by reference to instruments, from an instructional standpoint.
4. Analyzes and corrects simulated common errors related to straight, constant airspeed climbs, solely by reference to instruments.

C. TASK: CONSTANT AIRSPEED DESCENTS (AMEL and AMES)

REFERENCES FAA-H-8083-3, FAA-H-8083-15; FAA-S-8081-14.

Objective. To determine that the applicant:

1. Exhibits instructional knowledge of the elements of straight and turning, constant airspeed descents, solely by reference to instruments by describing—

   a. instrument crosscheck, instrument interpretation, and aircraft control.
   b. instruments used for pitch, bank, and power control during entry, during the descent, and during level off, and how those instruments are used to maintain descent heading and airspeed.
   c. trim technique.

2. Exhibits instructional knowledge of common errors related to straight and turning, constant airspeed descents, solely by reference to instruments by describing—

   a. "fixation," "omission," and "emphasis" errors during instrument crosscheck.
   b. improper instrument interpretation.
   c. improper control applications.
   d. failure to establish proper pitch, bank, or power adjustments during heading and airspeed corrections.
   e. improper entry or level-off technique.
   f. faulty trim technique.

3. Demonstrates and simultaneously explains a straight, constant airspeed descent, solely by reference to instruments, from an instructional standpoint.
4. Analyzes and corrects simulated common errors related to straight, constant airspeed descents, solely by reference to instruments.

D. TASK: TURNS TO HEADINGS (AMEL and AMES)

REFERENCES FAA-H-8083-3, FAA-H-8083-15; FAA-S-8081-14.

Objective. To determine that the applicant:

1. Exhibits instructional knowledge of the elements of turns to headings, solely by reference to instruments by describing—
   a. instrument crosscheck, instrument interpretation, and aircraft control.
   b. instruments used for pitch, bank, and power control during turn entry, during the turn, and during the turn rollout, and how those instruments are used.
   c. trim technique.
2. Exhibits instructional knowledge of common errors related to turns to headings, solely by reference to instruments by describing—
   a. "fixation," "omission," and "emphasis" errors during instrument crosscheck.
   b. improper instrument interpretation.
   c. improper control applications.
   d. failure to establish proper pitch, bank, and power adjustments during altitude, bank, and airspeed corrections.
   e. improper entry or rollout technique.
   f. faulty trim technique.
3. Demonstrates and simultaneously explains a turn to a heading, solely by reference to instruments, from an instructional standpoint.
4. Analyzes and corrects simulated common errors related to turns to headings, solely by reference to instruments.

E. TASK: RECOVERY FROM UNUSUAL FLIGHT ATTITUDES
(AMEL and AMES)

REFERENCES: FAA-H-8083-3, FAA-H-8083-15; FAA-S-8081-14.

Objective. To determine that the applicant:

1. Exhibits instructional knowledge of the elements of recovery from unusual flight attitudes by describing—
   a. conditions and situations that may result in unusual flight attitudes.
   b. the two basic unusual flight attitudes—nose-high (climbing turn) and nose-low (diving spiral).
   c. how unusual flight attitudes are recognized.
   d. control sequence for recovery from a nose-high attitude and the reasons for that sequence.
   e. control sequence for recovery from a nose-low attitude and the reasons for that sequence.
   f. reasons why the controls should be coordinated during unusual flight attitude recoveries.
2. Exhibits instructional knowledge of common errors related to recovery from unusual flight attitudes by describing—
   a. failure to recognize an unusual flight attitude.
   b. consequences of attempting to recover from an unusual flight attitude by "feel" rather than by instrument indications.
   c. inappropriate control applications during recovery.
   d. failure to recognize from instrument indications when the airplane is passing through a level flight attitude.
3. Demonstrates and simultaneously explains recovery from a nose-high and a nose-low unusual flight attitude from an instructional standpoint.
4. Analyzes and corrects simulated common errors related to recovery from unusual flight attitudes.

XIII. AREA OF OPERATION: EMERGENCY OPERATIONS

NOTE: The examiner shall select TASKs B or C, D, and one other TASK.

A. TASK: SYSTEMS AND EQUIPMENT MALFUNCTIONS
(AMEL and AMES)

REFERENCES: FAA-H-8083-3; FAA-S-8081-12, FAA-S-8081-14; POH/AFM.

NOTE: The examiner shall not simulate a system or equipment malfunction in a manner that may jeopardize safe flight or result in possible damage to the airplane.

Objective. To determine that the applicant exhibits instructional knowledge of the elements related to systems and equipment malfunctions, appropriate to the airplane used for the practical test, by describing recommended pilot action for at least of five (5) of the following:

1. Smoke, fire, or both, during ground or flight operations.
2. Rough running engine, partial power loss, or sudden engine stoppage.
3. Propeller malfunction.
4. Loss of engine oil pressure.
5. Fuel starvation.
6. Engine overheat.
7. Hydraulic system malfunction.
8. Electrical system malfunction.
9. Carburetor or induction icing.
10. Door or window opening in flight.
11. Inoperative or "runaway" trim.
12. Landing gear or flap malfunction.
13. Pressurization malfunction.

B. TASK: ENGINE FAILURE DURING TAKEOFF BEFORE V_{MC} (AMEL and AMES)

NOTE: Engine failure shall not be simulated at a speed greater than 50 percent V_{MC}.

REFERENCES: FAA-H-8083-3; FAA-S-8081-12, FAA-S-8081-14; POH/AFM.

Objective. To determine that the applicant:

1. Exhibits instructional knowledge of the elements related to engine failure during takeoff before V_{MC} by describing—
   a. use of prescribed emergency procedure.
   b. prompt closing of throttles.
   c. how to maintain directional control.
   d. proper use of brakes (landplane).
2. Exhibits instructional knowledge of common errors related to engine failure during takeoff before V_{MC} by describing—
   a. failure to follow prescribed emergency procedure.
   b. failure to promptly recognize engine failure.
   c. failure to promptly close throttles following engine failure.
   d. faulty directional control and use of brakes.
3. Demonstrates and simultaneously explains a simulated engine failure during takeoff before V_{MC} from an instructional standpoint.
4. Analyzes and corrects simulated common errors related to engine failure during takeoff before V_{MC}.

C. TASK: ENGINE FAILURE AFTER LIFT-OFF (AMEL and AMES)

REFERENCES: FAA-H-8083-3; FAA-S-8081-12, FAA-S-8081-14; POH/AFM.

Objective. To determine that the applicant:

1. Exhibits instructional knowledge of the elements related to engine failure after lift-off by describing—

    a. use of prescribed emergency checklist to verify accomplishment of procedures for securing the inoperative engine.
    b. proper adjustment of engine controls, reduction of drag, and identification and verification of the inoperative engine.
    c. how to establish and maintain a pitch attitude that will result in the best engine inoperative airspeed, considering the height of obstructions.
    d. how to establish and maintain a bank as required for best performance.
    e. how to maintain directional control.
    f. methods to be used for determining reason for malfunction.
    g. monitoring and proper use of the operating engine.
    h. an emergency approach and landing, if a climb or level flight is not within the airplane's performance capability.
    i. positive airplane control.
    j. how to obtain assistance from the appropriate facility.

2. Exhibits instructional knowledge of common errors related to engine failure after lift-off by describing—

    a. failure to follow prescribed emergency checklist.
    b. failure to properly identify and verify the inoperative engine.
    c. failure to properly adjust engine controls and reduce drag.
    d. failure to maintain directional control.
    e. failure to establish and maintain a pitch attitude that will result in best engine inoperative airspeed, considering the height of obstructions.
    f. failure to establish and maintain proper bank for best performance.

3. Demonstrates and simultaneously explains a simulated engine failure after lift-off from an instructional standpoint.
4. Analyzes and corrects simulated common errors related to engine failure after lift-off.

D. TASK: APPROACH AND LANDING WITH AN INOPERATIVE ENGINE (AMEL and AMES)

REFERENCES: FAA-H-8083-3; FAA-S-8081-12, FAA-S-8081-14; POH/AFM.

Objective. To determine that the applicant:

1. Exhibits instructional knowledge of the elements related to an approach and landing with an inoperative engine by describing—

    a. use of the prescribed emergency checklist to verify accomplishment of procedures for securing the inoperative engine.
    b. proper adjustment of engine controls, reduction of drag, and identification and verification of the inoperative engine.
    c. how to establish and maintain best engine inoperative airspeed.
    d. trim procedure.
    e. how to establish and maintain a bank as required for best performance.
    f. the monitoring and adjusting of the operating engine.
    g. proper approach to selected touchdown area, at the recommended airspeed.
    h. proper application of flight controls.
    i. how to maintain a precise ground track.
    j. wind shear and turbulence.
    k. proper timing, judgment, and control procedure during roundout and touchdown.
    l. directional control after touchdown.
    m. use of brakes (landplane).

2. Exhibits instructional knowledge of common errors related to an approach and landing with an inoperative engine by describing—

    a. failure to follow prescribed emergency checklist.
    b. failure to properly identify and verify the inoperative engine.
    c. failure to properly adjust engine controls and reduce drag.
    d. failure to establish and maintain best engine inoperative airspeed.
    e. improper trim procedure.
    f. failure to establish proper approach and landing configuration at appropriate time and in proper sequence.
    g. failure to use proper procedures for wind shear or turbulence.
    h. faulty technique during roundout and touchdown.
    i. improper directional control after touchdown.
    j. improper use of brakes (landplane).

3. Demonstrates and simultaneously explains an approach and landing with a simulated inoperative engine from an instructional standpoint.
4. Analyzes and corrects simulated common errors related to an approach and landing with an inoperative engine.

E. TASK: EMERGENCY DESCENT (AMEL and AMES)

REFERENCES: FAA-H-8083-3; FAA-S-8081-12, FAA-S-8081-14; POH/AFM

Objective: To determine that the applicant exhibits instructional knowledge of the elements related to emergency descents appropriate to the airplane flown by describing:

1. Exhibits instructional knowledge of the elements related to an emergency descent by describing—

   a. situations that require an emergency descent.
   b. proper use of the prescribed emergency checklist to verify accomplishment of procedures for initiating the emergency descent.
   c. proper use of clearing procedures before initiating and during the emergency descent.
   d. procedures for recovering from an emergency descent.
   e. manufacturer's procedures.

2. Exhibits instructional knowledge of common errors related to an emergency descent by describing—

   a. the consequences of failing to identify reason for executing an emergency descent.
   b. improper use of the prescribed emergency checklist to verify accomplishment of procedures for initiating the emergency descent.
   c. improper use of clearing procedures before initiating and during the emergency descent.
   d. improper procedures for recovering from an emergency descent.

3. Demonstrates and simultaneously explains an approach and landing with a simulated inoperative engine from an instructional standpoint.
4. Analyzes and corrects simulated common errors related to an approach and landing with an inoperative engine.

F. TASK: EMERGENCY EQUIPMENT AND SURVIVAL GEAR (AMEL and AMES)

REFERENCES: FAA-H-8083-3, FAA-H-8083-23; FAA-S-8081-12, FAA-S-8081-14; POH/AFM.

Objective. To determine that the applicant exhibits instructional knowledge of the elements related to emergency equipment and survival gear appropriate to the airplane flown by describing:

1. Equipment and gear appropriate for operation in various climates, over various types of terrain, and over water.
2. Purpose, method of operation or use, servicing, and storage of appropriate equipment.

XIV. AREA OF OPERATION: MULTIENGINE OPERATIONS

NOTE: The examiner shall select TASK D, E, and one other TASK.

A. TASK: OPERATION OF SYSTEMS (AMEL and AMES)

REFERENCES: FAA-H-8083-3; FAA-S-8081-12, FAA-S-8081-14; POH/AFM.

Objective. To determine that the applicant exhibits instructional knowledge of the elements related to the operation of systems, as applicable to the multiengine airplane used for the practical test, by describing at least five (5) of the following systems:

1. Primary flight controls and trim.
2. Flaps, leading edge devices, and spoilers.
3. Water rudders.
4. Powerplant and propellers.
5. Landing gear.
6. Fuel, oil, and hydraulic system.
7. Electrical.
8. Avionics.
9. Pitot static/vacuum system and associated instruments.
10. Environmental.
11. Deicing and anti-icing.

B. TASK: PERFORMANCE AND LIMITATIONS (AMEL and AMES)

REFERENCES: FAA-H-8083-3; FAA-S-8081-12, FAA-S-8081-14; POH/AFM.

Objective. To determine that the applicant exhibits instructional knowledge of the elements related to multiengine performance and limitations by describing:

1. Determination of weight and balance condition.
2. Use of performance charts, tables, and other data in determining performance in various phases of flight.
3. Effects of exceeding limitations.
4. Effects of atmospheric conditions on performance.
5. Factors to be considered in determining that the required performance is within the airplane's single and multiengine capabilities.

C. TASK: FLIGHT PRINCIPLES—ENGINE INOPERATIVE
(AMEL and AMES)

REFERENCES: FAA-H-8083-3; FAA-S-8081-12, FAA-S-8081-14; POH/AFM.

Objective. To determine that the applicant exhibits instructional knowledge of the elements related to flight principles—engine inoperative by describing:

1. Meaning of the term "critical engine."
2. Effects of density altitude on the V_{MC} demonstration.
3. Effects of airplane weight and center of gravity on control.
4. Effects of bank angle on V_{MC}.
5. Relationship of V_{MC} to stall speed.
6. Reasons for loss of directional control.
7. Indications of loss of directional control.
8. Importance of maintaining the proper pitch and bank attitude, and the proper coordination of controls.
9. Loss of directional control recovery procedures.
10. Engine failure during takeoff including planning, decisions, and single-engine operations.

D. TASK: MANEUVERING WITH ONE ENGINE INOPERATIVE (AMEL and AMES)

REFERENCES: FAA-H-8083-3; FAA-S-8081-12, FAA-S-8081-14; POH/AFM.

NOTE: The feathering of one propeller shall be demonstrated in any multiengine airplane equipped with propellers, which can be safely feathered and unfeathered in flight. Feathering for pilot flight test purposes should be performed only under such conditions and at such altitudes (no lower than 3,000 feet/1,000 meters above the surface) and positions where safe landings on established airports can be readily accomplished, in the event difficulty is encountered in unfeathering. At altitudes lower than 3,000 feet above the surface, simulated engine failure will be performed by throttling the engine and then establishing zero thrust.

In the event a propeller cannot be unfeathered during the practical test, it should be treated as an emergency.

Objective. To determine that the applicant:

1. Exhibits instructional knowledge of the elements related to maneuvering with one engine inoperative by describing—

    a. flight characteristics and controllability associated with maneuvering with one engine inoperative.
    b. use of prescribed emergency checklist to verify accomplishment of procedures for securing inoperative engine.
    c. proper adjustment of engine controls, reduction of drag, and identification and verification of the inoperative engine.
    d. how to establish and maintain the best engine inoperative airspeed.
    e. proper trim procedure.
    f. how to establish and maintain a bank, as required, for best performance.
    g. appropriate methods to be used for determining the reason for the malfunction.
    h. importance of establishing a heading toward the nearest suitable airport or seaplane base.
    i. importance of monitoring and adjusting the operating engine.
    j. performance of straight-and-level flight, turns, descents, and climbs, if the airplane is capable of those maneuvers under existing conditions.

2. Exhibits instructional knowledge of common errors related to maneuvering with one engine inoperative by describing—

    a. failure to follow prescribed emergency checklist.
    b. failure to recognize an inoperative engine.
    c. hazards of improperly identifying and verifying the inoperative engine.
    d. failure to properly adjust engine controls and reduce drag.
    e. failure to establish and maintain the best engine inoperative airspeed.
    f. improper trim procedure.
    g. failure to establish and maintain proper bank for best performance.
    h. failure to maintain positive control while maneuvering.
    i. hazards of attempting flight contrary to the airplane's operating limitations.

3. Demonstrates and simultaneously explains maneuvering with one engine inoperative from an instructional standpoint.
4. Analyzes and corrects simulated common errors related to maneuvering with one engine inoperative.

E. TASK: V_{MC} DEMONSTRATION (AMEL and AMES)

REFERENCES: FAA-H-8083-3; FAA-S-8081-12, FAA-S-8081-14; POH/AFM.

NOTE: Performing this maneuver by increasing pitch attitude to a high angle with both engines operating and then reducing power on the critical engine should be avoided. This technique is hazardous and may result in loss of aircraft control.

Objective. To determine that the applicant:

1. Exhibits instructional knowledge of the elements related to V_{MC} demonstration by describing—

   a. causes of loss of directional control at airspeeds less than V_{MC}, the factors affecting V_{MC}, and the safe recovery procedures.
   b. establishment of airplane configuration, adjustment of power controls, and trim prior to the demonstration.
   c. establishment of engine inoperative pitch attitude and airspeed.
   d. establishment of a bank attitude as required for best performance.
   e. entry procedure to demonstrate loss of directional control.
   f. indications that enable a pilot to recognize loss of directional control.
   g. proper recovery procedure.

2. Exhibits instructional knowledge of common errors related to V_{MC} demonstration by describing—

   a. inadequate knowledge of the causes of loss of directional control at airspeeds less than V_{MC}, factors affecting V_{MC}, and safe recovery procedures.
   b. improper entry procedures, including pitch attitude, bank attitude, and airspeed.
   c. failure to recognize imminent loss of directional control.
   d. failure to use proper recovery procedure.

3. Demonstrates and simultaneously explains engine inoperative loss of directional control from an instructional standpoint.
4. Analyzes and corrects simulated common errors related to engine inoperative loss of directional control.

F. TASK: DEMONSTRATING THE EFFECTS OF VARIOUS AIRSPEEDS AND CONFIGURATIONS DURING ENGINE INOPERATIVE PERFORMANCE (AMEL and AMES)

REFERENCES: FAA-H-8083-3; FAA-S-8081-12, FAA-S-8081-14; POH/AFM.

Objective. To determine that the applicant:

1. Exhibits instructional knowledge of the elements related to the effects of various airspeeds and configurations during engine inoperative performance by describing—

   a. selection of proper altitude for the demonstration.
   b. proper entry procedure to include pitch attitude, bank attitude, and airspeed.
   c. effects on performance of airspeed changes at, above, and below V_{YSE}.
   d. effects on performance of various configurations—

      (1) extension of landing gear.
      (2) extension of wing flaps.
      (3) extension of both landing gear and wing flaps.
      (4) windmilling of propeller on inoperative engine.

   e. airspeed control throughout the demonstration.
   f. proper control technique and procedures throughout the demonstration.

2. Exhibits instructional knowledge of common errors related to the effects of various airspeeds and configurations during engine inoperative performance by describing—

   a. inadequate knowledge of the effects of airspeeds above or below V_{YSE} and of various configurations on performance.
   b. improper entry procedures, including pitch attitude, bank attitude, and airspeed.
   c. improper airspeed control throughout the demonstration.
   d. rough and/or uncoordinated control technique.
   e. improper procedures during resumption of cruise flight.

3. Demonstrates and simultaneously explains the effects of various airspeeds and configurations during engine inoperative performance from an instructional standpoint.
4. Analyzes and corrects simulated common errors related to the effects of various airspeeds and configurations during engine inoperative performance.

XV. AREA OF OPERATION: POSTFLIGHT PROCEDURES

NOTE: The examiner shall select TASK A and for ASES applicants at least one other TASK.

A. TASK: POSTFLIGHT PROCEDURES (AMEL and AMES)

REFERENCES: FAA-H-8083-3, FAA-H-8083-23; FAA-S-8081-12, FAA-S-8081-14; POH/AFM.

Objective. To determine that the applicant:

1. Exhibits instructional knowledge of the elements of postflight procedures by describing—

    a. parking procedure (landplane).
    b. engine shutdown and securing cockpit.
    c. deplaning passengers.
    d. securing airplane.
    e. postflight inspection.
    f. refueling.

2. Exhibits instructional knowledge of common errors related to postflight procedures by describing—

    a. hazards resulting from failure to follow recommended procedures.
    b. poor planning, improper procedure, or faulty judgment in performance of postflight procedures.

3. Demonstrates and simultaneously explains postflight procedures from an instructional standpoint.
4. Analyzes and corrects simulated common errors related to postflight procedures.

B. TASK: ANCHORING (AMES)

REFERENCES: FAA-H-8083-3, FAA-H-8083-23; FAA-S-8081-12, FAA-S-8081-14; POH/AFM.

Objective. To determine that the applicant:

1. Exhibits instructional knowledge of the elements of anchoring by describing—
   a. how to select a suitable area for anchoring.
   b. recommended procedure for anchoring in a lake, river, or tidal area.
   c. number of anchors and lines to be used to ensure seaplane security in various conditions.
   d. hazards to be avoided during anchoring.
   e. requirements for anchoring lights.
2. Exhibits instructional knowledge of common errors related to anchoring by describing—
   a. hazards resulting from failure to follow recommended anchoring procedures.
   b. consequences of poor planning, improper procedure, or poor judgment when anchoring.
   c. consequences of failure to use anchoring lines of adequate length and strength to ensure seaplane security.
3. Demonstrates and simultaneously explains anchoring from an instructional standpoint.
4. Analyzes and corrects simulated common errors related to anchoring.

C. TASK: DOCKING AND MOORING (AMES)

REFERENCES: FAA-H-8083-3, FAA-H-8083-23; FAA-S-8081-12, FAA-S-8081-14; POH/AFM.

Objective. To determine that the applicant:

1. Exhibits instructional knowledge of the elements of docking and mooring by describing—
   a. how to select a suitable area for docking and mooring.
   b. recommended procedure for docking and mooring in a lake, river, or tidal area.
   c. number of tie-downs and lines to be used to ensure seaplane security in various conditions.
   d. hazards to be avoided during docking and mooring.
   e. requirements for docking and mooring lights.
2. Exhibits instructional knowledge of common errors related to docking and mooring by describing—
   a. hazards resulting from failure to follow recommended procedures.
   b. consequences of poor planning, improper procedure, or poor judgment when docking and mooring.
   c. consequences of failure to use docking and mooring lines of adequate length and strength to ensure seaplane security.
3. Demonstrates and simultaneously explains docking and mooring from an instructional standpoint.
4. Analyzes and corrects simulated common errors related to docking and mooring.

D. TASK: BEACHING (AMES)

REFERENCES: FAA-H-8083-3, FAA-H-8083-23; FAA-S-8081-12, FAA-S-8081-14; POH/AFM.

Objective. To determine that the applicant:

1. Exhibits instructional knowledge of the elements of beaching by describing—

    a. recommended procedures for beaching.
    b. factors to be considered such as beach selection, water depth, current, tide, and wind.
    c. procedures to be followed to ensure seaplane security.
    d. hazards to be avoided.

2. Exhibits instructional knowledge of common errors related to beaching by describing—

    a. hazards resulting from failure to follow recommended procedures.
    b. consequences of poor beach selection, poor planning, improper procedure, or faulty judgment when beaching.
    c. a consequence of failure to take appropriate precautions to avoid hazards or to ensure that seaplane is secure.

3. Demonstrates and simultaneously explains beaching from an instructional standpoint.
4. Analyzes and corrects simulated common errors related to beaching.

E. TASK: RAMPING (AMES)

REFERENCES: FAA-H-8083-3, FAA-H-8083-23; FAA-S-8081-12, FAA-S-8081-14; POH/AFM.

Objective. To determine that the applicant:

1. Exhibits instructional knowledge of the elements of ramping by describing—
    a. Factors, such as type of ramp surface, wind, current, and direction and control of approach speed.
    b. recommended procedures for ramping.
    c. hazards to be avoided.
2. Exhibits instructional knowledge of common errors related to ramping by describing—
    a. hazards resulting from failure to follow recommended procedures.
    b. consequences of poor planning, improper procedure, or faulty judgment when ramping.
    c. consequences of failure to take appropriate precautions to avoid hazards or to ensure that the seaplane is secure.
3. Demonstrates and simultaneously explains ramping from an instructional standpoint.
4. Analyzes and corrects simulated common errors related to ramping.